Sweet Biochemistry

Remembering Structures, Cycles and Pathways by Mnemonics

Sweet Biochemistry

Remembering Structures, Cycles and Pathways by Mnemonics

Second Edition

Asha Kumari
Department of Biochemistry, Shaheed Hasan Khan Mewati Government Medical College, Nuh, Haryana, India

Academic Press is an imprint of Elsevier
125 London Wall, London EC2Y 5AS, United Kingdom
525 B Street, Suite 1650, San Diego, CA 92101, United States
50 Hampshire Street, 5th Floor, Cambridge, MA 02139, United States
The Boulevard, Langford Lane, Kidlington, Oxford OX5 1GB, United Kingdom

ISBN: 978-0-443-15348-8

For Information on all Academic Press publications
visit our website at https://www.elsevier.com/books-and-journals

Publisher: Stacy Masucci
Acquisitions Editor: Ali Afzal-Khan
Editorial Project Manager: Michaela Realiza
Production Project Manager: Fahmida Sultana
Cover Designer: Christian J. Bilbow

Typeset by MPS Limited, Chennai, India

Dedication

Dedicated to Lord Hanuman:
Budhdhiheen tanu janike,
Sumiron pawan kumar,
Bal budhdhi vidya dehun mohi,
Harhu kalesh vikar

Contents

List of figures

List of tables

Preface

Science is an ever-expanding passion of brilliant minds. For healing a patient, enormous amount of information has to be remembered by the physician. We memorise many things by reading books, publications, writing notes and answers, meeting and treating patients. At times, it becomes very difficult to recall the important signs and symptoms of myriad diseases as the names and features of syndromes are increasing exponentially due to the rapid advancement of technology. Therefore it is crucial to learn the diseases and mechanisms with something unique.

In the first edition of Sweet Biochemistry, I had presented new mnemonics, poems, stories and tricks to remember the difficult biochemistry topics along with brief traditional text recap. Students and teachers were amazed to see funny and unique representation of intricate pathways, structures and diseases. In the second edition, I have added the core concept of each chapter to supplement the fun component. The readers can therefore enjoy the stories, poems and mnemonics along with having a deeper understanding of topic. The addition of molecular biology topics such as DNA structure, replication, transcription, translation and immunology topics including antigen, antibody, class switching and vaccines will extend the learning experience.

One may wonder why a syndrome is depicted as a person. The mnemonics are not intended to make fun of anyone. These are just an attempt to make a way to remember what is most important for patient care. The treatment given by a doctor matters the most and not the trick of remembering. I hope the students, teachers and physicians will appreciate my sincere efforts to present each concept in a unique way. If we make similar mnemonics, we often confuse the derivations. Hence, the reading of text with full attention and then absorbing the mnemonic is needed.

This book is for you if are a medical student, teacher, researcher or physician because foundation of biochemistry is important in medical science. I have always tried to make the subject enjoyable so that we do not feel that biochemistry is monotonous and there is very less to see with your eyes. Actually, there is so much to imagine and learn. My only request is that you should read this book as a story book of biochemistry, cherishing each and every new expression of molecules.

Asha Kumari

Acknowledgements

I am really filled with gratitude for the readers of the first edition of Sweet Biochemistry who appreciated the concept of book. This is the sole aim of an author while writing any book. I would like to express my deepest appreciation to Mr. Peter Linsley, Editor, Elsevier Academic Press for giving me the opportunity to bring the second edition into existence. I enjoyed the pleasure of working with the Editorial Project Manager Michaela Realiza and Fahmida Sultana in this journey. My inspirational teacher Dr. Shashi Seth Mam has always taught me to learn the subject deeply, which I cannot miss to mention. I would like to acknowledge my colleagues from PGIMS, Rohtak – Dr. Manish Raj Kulshreshtha, Dr. Piyush Bansal, Dr. Pawan Mittal, Dr. Deepika, Dr. Shinky and Dr. Rajesh for their sincere encouragement.

This endeavour would not have been possible without the blessings of Lord Hanuman and my parents Mr. Dayanand and Mrs. Burfo Devi. My elder brother Mr. Ashok Kumar, his wife Mrs. Indu and kids Sanvi, Siddharth and Manasvi deserve my heartfelt gratitude which is difficult to be expressed in words for filling my life with dreams and happiness. Special thanks to brother Mr. Sonu and his family for always motivating me. I would like to recognise the affection received by my younger brother Mr. Arun Kumar.

I would be remiss if I do not mention my department and students of SHKM GMC, Nalhar, Nuh, who are a constant cause of my improvement.

Chapter 1

Glycolysis

Microview

This pathway is extremely ancient, and you must understand the reason for its existence. We always say that the sun is the fundamental support of life on earth because without its sunlight, plants cannot perform photosynthesis. Plant organelle known as chloroplasts utilises sunlight to combine carbondioxide with water to form carbohydrates (one of which is glucose). Now when we eat food containing carbohydrates, we want to get that trapped energy to run our own pathways. Therefore, we burn the complex molecules such as glucose into simpler smaller molecules. That is why in glycolysis (breakdown of glucose), the 6-carbon compound is divided into two 3-carbon compounds. During this lysis, energy is released and ADP is converted into ATP – the energy currency of cell.

Glycolysis is a cytoplasmic pathway which breaks Glucose in two 3-carbon compounds and generates energy. Glycolysis is used by all the cells of the body for energy production. Glucose is trapped by phosphorylation, with the help of Hexokinase enzyme. ATP is used in this reaction, and product Glucose-6-Phosphate(G-6-P) inhibits Hexokinanse. This is irreversible reaction. G-6-P is isomerised into its ketose form Fructose-6-Phosphate by Phosphohexose isomerase.

F-6-P is further phosphorylated by Phosphofructokinase to Fructose 1,6-bisphosphate. This reaction is irreversible and the principal regulatory step. Aldolase cleaves F1,6 bisP into Glyceraldehyde-3-P and DHAP which are interconverted by enzyme Phosphotriose isomerase. Glyceraldehyde-3-P is oxidised by NAD + -dependent dehydrogenase forming 1,3-bisphosphoglycerate.

1,3-Bisphosphoglycerate has a high-energy acyl-phosphate bond and carries out substrate-level phosphorylation generating ATP. Enzyme participating is Phosphoglycerate kinase, and 3-Phosphoglycerate is formed. Phosphoglycerate mutase isomerises 3-Phosphoglycerate to 2-phosphoglycerate. 2-Phosphoglycerate is dehydrated by enolase to form Phosphoenolpyruvate, second compound capable of substrate-level phosphorylation in glycolysis. Pyruvate kinase transfers Phosphate group of Phosphoenolpyruvate to ADP, and pyruvate is formed.

Pyruvate enters the Krebs cycle in aerobic conditions, and in anaerobic conditions, it forms lactate which helps in generation of NAD + for continuation of glycolysis. Pyruvate is converted to acetyl-CoA by Pyruvate

Sweet Biochemistry. DOI: https://doi.org/10.1016/B978-0-443-15348-8.00017-X

Dehydrogenase Complex which is an irreversible step. Pyruvate enters the Krebs cycle for further energy production (Fig. 1.1).

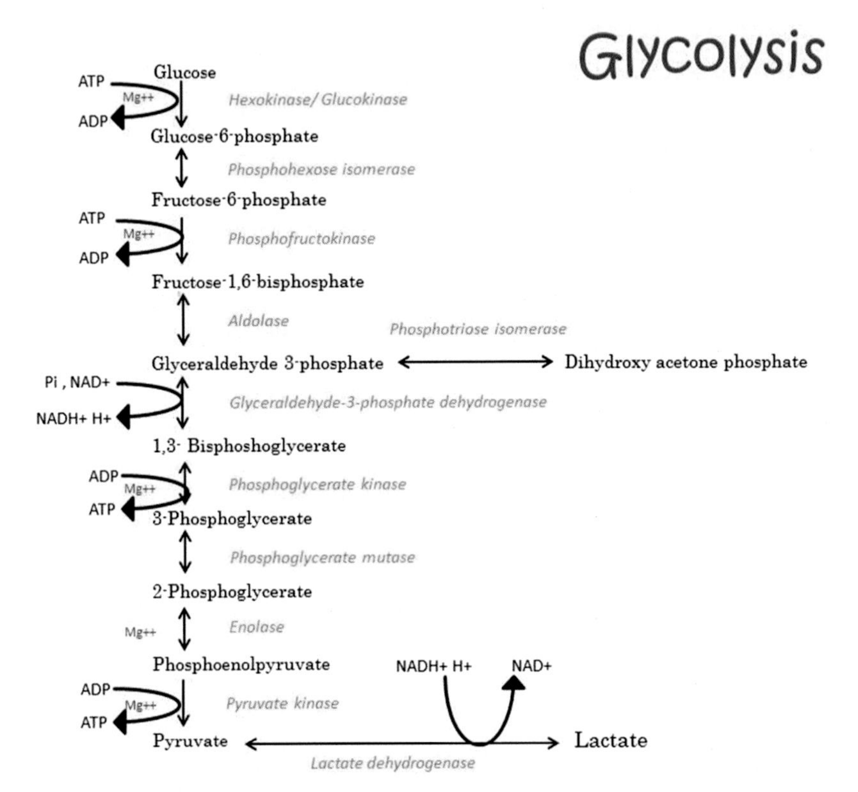

FIGURE 1.1 Glycolysis pathway.

Importance of glucose phosphorylation

The glucose enters blood circulation after the digestion and absorption of carbohydrates. This glucose molecule is enjoying the ride in blood stream but the cells are waiting for it. Cell have special gates, the 'Glucose Transporters' to facilitate the entry of glucose. Inside the cell, glucose is phosphorylated in the first step of glycolysis. But why a phosphate group is gifted to glucose on its entry? The negative charge of phosphate group prohibits the back-diffusion of glucose. So this is not a gift, it's a trap. The second effect is destabilising the glucose molecule. This is a common trick in pathways that any molecule which has to enter the reaction is first activated by addition of some charge.

Role of magnesium for kinases

The Hexokinase Enzyme which catalyses the phosphorylation of the Glucose is a kinase (and works as a glucose sensor secretly) and needs magnesium ion (Mg2 +) or some other divalent metal ion such as Mn2 + . Magnesium ion is not merely a witness of the reaction, rather Mg-ATP is the substrate of enzyme. Mg (with two positive charges) handles negative charges of ATP, thereby minimising unnecessary ionic interactions between ATP and enzyme. This facilitates the conformation recognised by the enzyme and by giving additional interaction site to the enzyme, increases the binding energy. The result of the first step of glycolysis, Glucose-6-phosphate is a free bird and can shuttle to other pathways such as HMP pathway and glycogen synthesis.

Critical isomerisation to fructose

Glucose-6-Phosphate, an Aldose, is converted to fructose-6-Phosphate, a ketose by Glucose-6-Phosphate isomerase (GPI). Glucose and fructose are predominantly cyclic structures. Hence, the GPI first opens the glucose-6-phosphate ring, then isomerises via enediol intermediate and finally closes the fructose-6-phosphate ring. This reaction can proceed in both directions quite easily. But actually it is one of the two reactions which are trying to make the unsymmetrical glucose-6-phosphate molecule symmetrical, for the fate of cleavage into two similar interconvertible three-carbon molecules. You can appreciate this statement from the figure (Fig. 1.2).

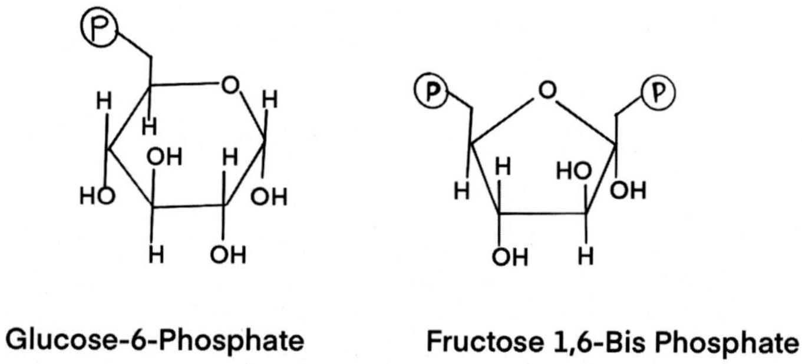

Glucose-6-Phosphate Fructose 1,6-Bis Phosphate

FIGURE 1.2 Structure of glucose-6-phosphate and fructose1,6 bisphosphate.

Second phosphorylation

Second phosphate is attached to the Fructose-6-phosphate by enzyme phosphofructokinase-1 synthesising Fructose 1,6-bisphosphate. For this reaction, an alcohol group is needed at C1 which is absent in G-6-P. By isomerisation of G-6-P to F-6-P, C2 gets ketone group while C1 receives alcohol group for phosphorylation. Presence of keto group at C2 is also prerequisite for the breakdown into two parts.

Fructose 1,6-bisphosphate is broken down into DHAP (a ketose) and Glyceraldehyde-3-phosphate (Gly-3-P an aldose) by enzyme Aldolase. The reaction is a reverse of Aldol condensation, hence the name is given.

Reason of DHAP conversion to Gly-3-P?

The product of aldolase at equilibrium is 96% of DHAP which participates in Triglyceride synthesis and only 4% Gly-3-P. But we observe in the pathway that it isomerises to Gly-3-P by the enzyme triose phosphate isomerase (TIM) through an enediol intermediate. To continue the reaction towards Gly-3-P formation, it is removed from the reaction site (by the next step of the pathway), so that more DHAP continues to transform and enter glycolysis (for ATP production), instead of joining triglyceride synthesis. This is known as Le Chatelier's principle.

Conversion of Gly-3-P to 1,3-bisphosphoglycerate

Gly-3-P (an aldehyde) is oxidised by NAD + in the presence of dehydrogenase to form a carboxylic acid which is phosphorylated by orthophosphate (and not ATP), generating a high-energy compound – 1,3-bisphosphoglycerate (an acyl-phosphate product). The energy of the oxidation step is conserved in the thioester intermediate and transformed into high phosphoryl-transfer potential which leads to substrate-level phosphorylation in the next step. This reaction becomes possible because the oxidation state is favourable and the phosphorylation is unfavourable. The coupling of these two reactions by thioester intermediate is very beneficial to ensure feasibility.

Reaction of phosphoglycerate kinase: the heroic step

For two previous phosphorylation steps, cell paid two ATPs. Now Phosphoglycerate kinase enzyme transfers a phosphoryl group from 1,3-bisphosphoglycerate to ADP, and the first ATP along with 3-phosphoglycerate is produced. You must remember that the glucose molecule was cut into two parts, and now both of them are entering reactions beyond aldolase catalysis. The ATP generation done here is called substrate-level phosphorylation because Electron transport chain is not involved here, and energy is directly used to phosphorylate ADP molecule. The source of energy mentioned here is carbon oxidation from previous step.

3-PG isomerises to 2-PG: the role

Phosphoglycerate mutase relocates phosphoryl group from 3-carbon to 2-carbon. The intramolecular shift of a group allocates the title mutase to the enzyme. However, the final phosphoryl group on C-2 is not the one removed from C-3 position.

The enzyme Phosphoglycerate mutase needs 2,3 bisphosphoglycerate as cofactor to keep a specific histidine residue at active site in phosphorylated form. At the active site, when 3-Phosphoglycerate arrives, the enzyme attaches the phosphate from its histidine to it, making 2,3-bisphosphoglycerate intermediate. Removal of phosphate from histidine induces a conformational change of enzyme, aligning the phosphate at C3 near its histidine at active site. Phosphate transfer to histidine is accompanied with 2-Phosphoglycerate release and returning of enzyme to its original state.

Importance of enolase reaction

Second last reaction of Glycolysis is dehydration of 2-Phosphoglycerate by enolase enzyme to form Phosphoenolpyruvate. The transfer potential of phosphoryl group is greatly increased by this reaction as the product, an enol pyruvate is a brought to a unstable enol form due to the phosphorylation. By releasing this energy, it settles to pyruvate, a more stable ketone molecule. The ADP combines with phosphate group using this energy. Enzyme Pyruvate kinase catalyses this irreversible transfer. After understanding Glycolysis pathway, let us memorise it by a simple mnemonic provided in Figs. 1.3 and 1.4.

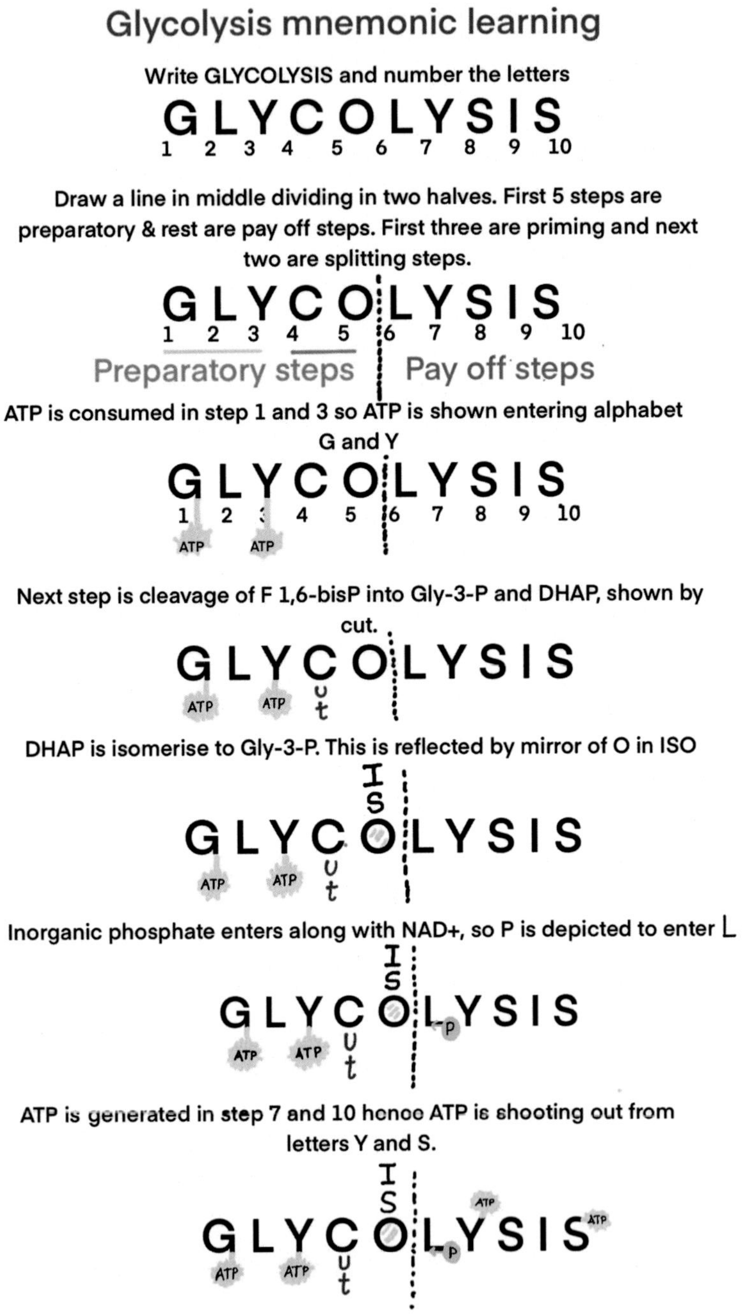

Glycolysis mnemonic learning

Write GLYCOLYSIS and number the letters

GLYCOLYSIS
1 2 3 4 5 6 7 8 9 10

Draw a line in middle dividing in two halves. First 5 steps are preparatory & rest are pay off steps. First three are priming and next two are splitting steps.

GLYCO | LYSIS
1 2 3 4 5 | 6 7 8 9 10
Preparatory steps | Pay off steps

ATP is consumed in step 1 and 3 so ATP is shown entering alphabet G and Y

GLYCO | LYSIS
1 2 3 4 5 | 6 7 8 9 10
ATP ATP

Next step is cleavage of F 1,6-bisP into Gly-3-P and DHAP, shown by cut.

GLYCO | LYSIS
ATP ATP Ut

DHAP is isomerise to Gly-3-P. This is reflected by mirror of O in ISO

IS
GLYCO | LYSIS
ATP ATP Ut

Inorganic phosphate enters along with NAD+, so P is depicted to enter L

IS
GLYCO | LPYSIS
ATP ATP Ut

ATP is generated in step 7 and 10 hence ATP is shooting out from letters Y and S.

IS ATP
GLYCO | LPYSIS ATP
ATP ATP Ut

FIGURE 1.3 Mnemonic for glycolysis, part-1.

Eighth step is isomerisation in which molecular formula is SAME and only the phosphate position is changed.

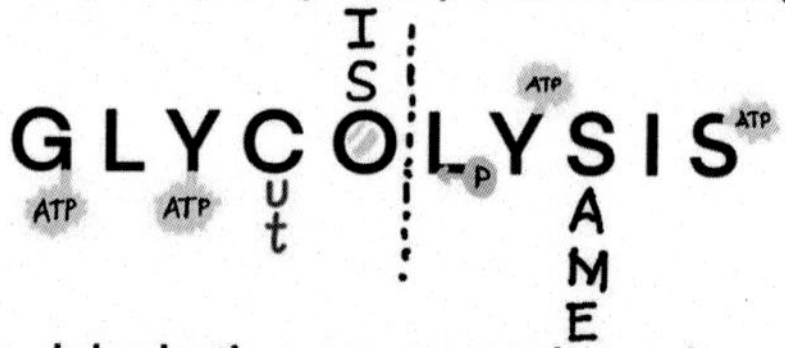

In ninth step, dehydration occurs as shown by water coming out form pipe.

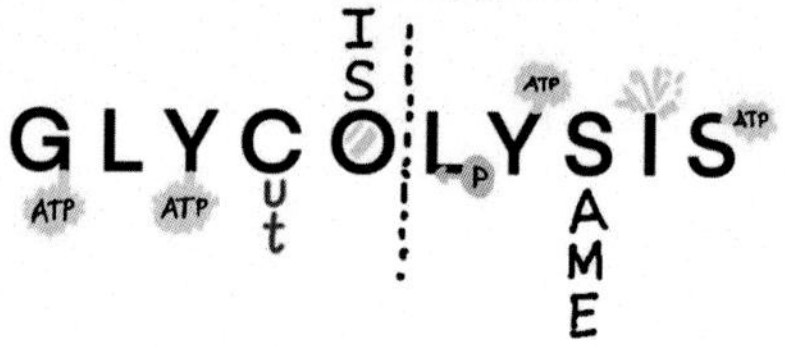

To remember the names of enzymes involved in the steps, remember that all enzymes except last two are named after the substrate. Enolase and Pyruvate Kinase are named after the products.

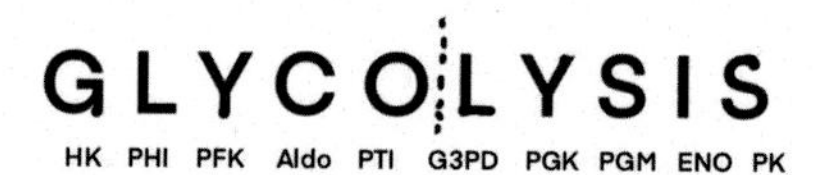

Enzymes HK- Hexokinase, PHI- Phospoexose isomerase, PFK- Phosphofructokinase, Aldo- Aldolase, PTI- Phosphotriose isomerase, G-3-PD Glyceraldehyde-3-Phsophate dehydrogenase, PGK- Phosphoglycerate kinase, PGM- Phosphoglycerate mutase, Eno- Enolase, PK- Pyruvate Kinase..

FIGURE 1.4 Mnemonic for glycolysis, part-2.

Chapter 2

Citric acid cycle

All of us have used wood in making fire for cooking food or bonfire. We like to burn the wood to the completion. If wood is half burnt, extraction of energy from it is incomplete. We again burn it till ash is formed. Lot of smoke including carbondioxide is released in the process along with heat. This heat is used by us.

What we call fuels for our body are the carbon compounds which are oxidisable. Process such as glycolysis cannot fully extract the energy out of the major fuel carbohydrates (in the form of adenosine triphosphate (ATP) or high-energy electrons), so the end product of glycolysis – Pyruvate is channelised into tricarboxylic acid (TCA) cycle for further oxidation-reduction reactions. Proteins and fats which can convert into acetyl-coenzyme A (CoA) also enter Krebs cycle (or Tricarboxylic acid cycle/TCA cycle) to provide more energy. The highlight of this cycle is oxidative decarboxylation reactions which release carbon atoms of acetyl-CoA as CO_2 (to be precise, the atoms released as CO_2 are not original atoms entering as acetyl-CoA) and water.

You may wonder why this pathway is called a cycle unlike glycolysis or many others. The reason is that the final derivative here generates the substrate of first reaction. Last creates first and first follows till last. Therefore, if the concentration of substrates is adequate, the cycle can go relentlessly.

This cycle is amphibolic as molecules are not only oxidised, but some intermediates can participate in the synthesis of amino acids, nucleotide bases, porphyrin ring and glucose. The intermediate molecules are not consumed in the reactions and are regenerated after one cycle (intermediates are behaving like catalysts).

Quick view of pathway

The Citric acid cycle uses mitochondrial enzymes for final oxidation of carbohydrates, proteins and fats. Moreover, Krebs cycle also produces intermediates which are important in gluconeogenesis, lipolysis, neurotransmitter synthesis, etc. Acetyl-CoA from pyruvate of glycolysis, beta oxidation of fatty acids, ketogenic amino acids and ketones enters this pathway for energy production. First step is fusion of acetyl group of acetyl-CoA with oxaloacetate, catalysed by Citrate Synthase. CoA-SH and heat are released, and

Sweet Biochemistry. DOI: https://doi.org/10.1016/B978-0-443-15348-8.00007-7

Citrate is the product. Citrate is isomerised by dehydration and rehydration to isocitrate. Aconitase enzyme catalyses these two steps via cis-aconitate as intermediate. Next two steps are catalysed by Isocitrate dehydrogenase. Dehydrogenation of Isocitrate forms Oxalosuccinate which decarboxylates to alpha-Ketoglutarate. Alpha-Ketoglutarate is further oxidatively decarboxylated by alpha-ketoglutarate dehydrogenase – a multienzyme complex. Succinyl-CoA is formed in this unidirectional reaction.

Succinate Thiokinase converts Succinyl-CoA to Succinate, meanwhile generating first ATP/guanosine triphosphate (GTP) by substrate-level phosphorylation. Succinate is acted upon by Succinate Dehydrogenase requiring FAD and Fe-S proteins to form Fumarate. Fumarase adds water on double bond of Fumarate, and malate is yielded. Malate regenerates Oxaloacetate by action of NAD + -dependent Malate dehydrogenase, completing the cycle.

Coenzymes such as FAD and NAD + are reduced in Krebs cycle, which transfers electrons to electron transport chain for final oxygen as acceptor. Three nicotinamide adenine dinucleotide (NADH +) and one FADH2 are generated in one cycle which on entering electron transport chain yields 10 ATP. These include one ATP produced by Succinate thiokinase at substrate level. Two carbon atoms are lost in this cycle by decarboxylation, although these are not the same atoms entering as Acetyl-CoA. Vitamins such as Riboflavin, Niacin and Thiamine work as coenzyme in this cycle while Pantothenic acid forms CoA part of acetyl-CoA.

Fluoroacetate inhibits aconitase enzyme, arsenite inhibits alpha-ketoglutarate and malonate iinhibits succinate dehydrogenase enzyme.

Krebs cycle takes place in the liver primarily. In case of enzyme defects of Krebs cycle, ATP production is hampered to a great extent, leading to severe brain damage (Fig. 2.1).

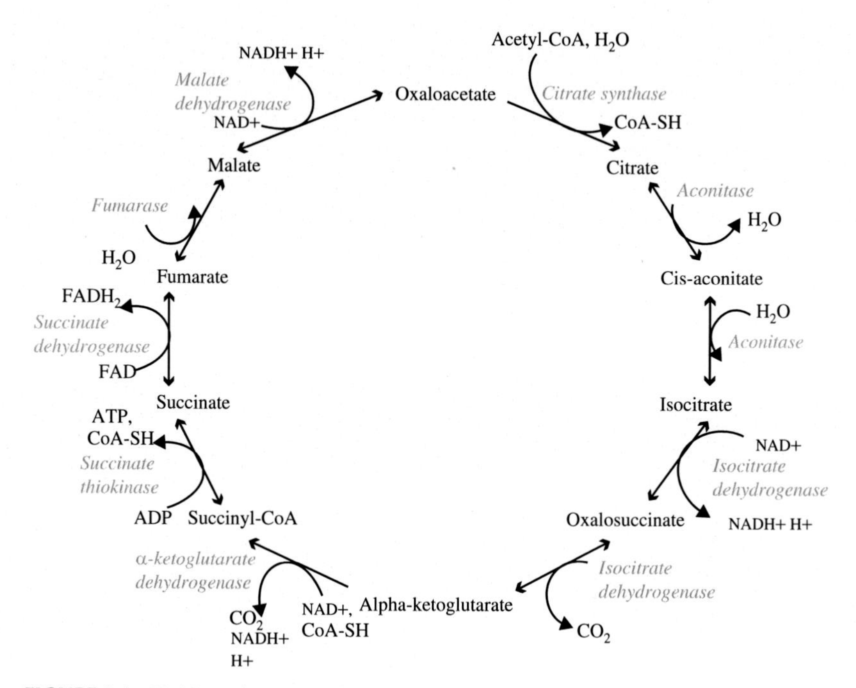

FIGURE 2.1 Kreb's cycle.

Reactions in detail

First reaction of TCA cycle is condensation of acetyl-CoA (having two carbons) with 4-carbon molecule oxaloacetate. Condensation occurs between carbanion formed at methyl group of acetyl-CoA and carbonyl group of oxaloacetate to form citrate. This aldol condensation is irreversible being an exergonic reaction. The acetyl-CoA, which is qualified enough to enter other pathways, for example, fatty acid synthesis, ketone body synthesis and acetylcholine synthesis becomes committed after this step conducted in the mitochondria. The enzyme Citrate synthase working here needs initial binding of oxaloacetate to bring a conformational change in it, allowing binding of second substrate acetyl-CoA.

Do you want to know the purpose of acetyl-CoA formation?

This is really interesting. In eukaryotes, during aerobic respiration, Pyruvate generated at the end of glycolysis is sent to the mitochondria (by a recently discovered mitochondrial Pyruvate carrier) where externally supplied oxygen accepts the high-energy electrons. For this, more oxidation and decarboxylation are done step by step till every carbon of Pyruvate is released as carbon-dioxide. Acetyl-CoA is formed by Pyruvate dehydrogenase complex which catalyses this oxidation + decarboxylation, along with forming thioester linkage. This bond links acetyl group with sulfhydryl substituent of beta-mercaptoethylamine group and possess high energy. Therefore, the hydrolysis of thioester bond generates large amount of energy.

Role of citrate's isomerisation to isocitrate

Citrate is an allosteric regulator of fatty acid synthesis (positive modifier) and glycolysis (negative modifier). It is acted upon by enzyme Aconitate dehydrates/Aconitase, having four iron–four sulphur centre. Citrate is a symmetric molecule, but the Aconitase intelligently selects CH2COOH group coming from oxaloacetate by stereospecific three-point attachment. Aconitase first removes water molecule and then adds rehydrates, causing the isomerisation. This is important because secondary alcoholic group (present in isocitrate) is easier to oxidise than tertiary alcoholic group of citrate.

Action of isocitrate dehydrogenase enzyme: first oxidative decarboxylation

Isocitrate dehydrogenase (ICD) requires magnesium or manganese as cofactor and has three isoenzyme forms. It catalyses conversion of isocitrate (six carbon molecules) to alpha-ketoglutarate. The latter has one carbon less. This reaction takes place in two parts. As the name suggests, first is oxidation to form beta-keto acid oxalosuccinate, and second is decarboxylation to form alpha-ketoglutarate. Oxidation is linked to reduction of NAD + or NADP + .

Second oxidative decarboxylation and adddition of CoA

Alpha-ketoglutarate (KG) dehydrogenase is a multienzyme complex having multiple copies of three enzymes – Alpha-KG dehydrogenase, dihydrolipoyl transsucciylase and dihydrolipoyl dehydrogenase. This big molecular machine resembles Pyruvate dehydrogenase complex (PDC) in functioning and requirements. However, PDC is finely regulated by a kinase and a phosphatase while this enzyme is not.

Alpha-KG, an alpha keto acid by oxidation and decarboxylation generates a thioester – with one carbon less. NAD + is also reduced at this step. Alpha-KG can participate in transamination to convert into a neurotransmitter glutamate. Another role of alpha-KG is post-translational hydroxylation of collagen at prolyl or lysyl residues.

CoA removal from succinyl-CoA coupled with GTP formation

As mentioned earlier, hydrolysis of thioester linkage releases lot of energy. Succinate thiokinase separates Coenzyme-A from succinyl-CoA, and the energy of the reaction is captured in producing high-energy phosphoric anhydride bond which is called substrate-level phosphorylation of guanosine diphosphate (GDP) with phosphate along with product Succinate. This is a reversible reaction with a standard free energy of −0.8 kcal/mole. You may think why GDP is participating here. Why not adenosine diphosphate (ADP)? Actually either can be produced depending on the tissue. There are two isoenzymes of this enzyme. Isoenzyme present in the heart and skeletal muscle produces ATP while that present in the liver uses GDP. GTP formed can be transformed to ATP by the enzyme nucleoside diphosphate kinase.

A unique dehydrogenation of succinate to fumarate using FAD

In previous oxidations, NAD + or NADP + received electrons, but in this step, two hydrogens are accepted by flavin adenine dinucleotide (FAD +), forming FADH2. Now why this difference? The reason lies in the reduction potential of coupling two half reactions. The difference in reduction potential between the fumarate/Succinate and NAD + /NADH half reactions is not adequate to let NAD + participate in the reaction. This requirement is fulfilled by FAD + /FADH2 half reaction. Another point to note here is that the NADH is a soluble molecule, but FADH2 remains bound with enzyme Succinate dehydrogenase and transfers the electrons to electron transport chain directly at the coenzyme Q level because the enzyme is bound to the inner mitochondrial membrane.

Succinate dehydrogenase is a stereospecific enzyme which prefers acting on trans hydrogen atom of the methylene carbons of Succinate, forming Fumarate, a trans unsaturated dicarboxylate.

Fumarate to malate: a stereospecific hydration

Fumarase, also called Fumarate hydratase, performs a reversible trans addition of H and OH across the double bond of fumarate to form L-malate. The enzyme is stereospecific just like the previous step enzyme.

Malate dehydrogenation regenerates oxaloacetate, the substrate of first step

NAD + -linked malate dehydrogenase catalyses the dehydrogenation of L-malate. This oxidation yields oxaloacetate. The malate is really not interested in this conversion, but the consumption of oxaloacetate in condensation reaction by Citrate synthase removes it from the vicinity, therefore favouring forward reaction.

After this deep discussion, let me present a cute poem to remember this intricate pathway (Figs. 2.2 and 2.3).

FIGURE 2.2 Mnemonic poem for Kreb's cycle.

What relates to the Kreb's cycle intermediates in poem

A Cow & Ox were	Acetyl-CoA oxaloacetate
Citing, citing (read as seating)	Citrate
Cis AC "ON" on	Cis-Aconitate
Ceiling, Ceiling	
I C Ox Said	Iso-citrate, oxalosuccinate
Dancing, Dancing	
Al Ketl are now	Alpha-ketoglutarate
Sucking, sucking	Succinyl-CoA
Cold drinks and some	
Singing, singing	Succinate
Fine Maa, Maa	Fumarate, Malate
In Oxy ringing ringing	Oxaloacetate

FIGURE 2.3 The links of Poem with the molecules of Kreb's cycle.

Chapter 3

Electron transport chain

Electron transport chain (ETC) is a mitochondrial pathway in which electrons move across a redox span of 1.1 V from NAD + /NADH to O_2/H_2O. Three complexes are involved in this chain, namely complex I, complex III and complex IV. Some compounds such as succinate which have more positive redox potential than NAD + /NADH can transfer electrons via a different complex – complex II.

Coenzyme Q or simply Q can travel within the membrane while Cyt C is a soluble protein. Flavoproteins are components of complex I and II, and Fe-S is present in complex I, II and III. Fe atom present in Fe-S complexes helps in electron transfer by shifting from Fe^{2+} to Fe^{3+} states.

Electrons are transferred from NADH to Flavin mononucleotide (FMN), from where they enter Fe-S complexes. From Fe-S they move to Q which carries them to complex III. In complex III electrons are received by cytochrome c1 and cytochrome b and sent to Cyt c.

Cyt c transfers electrons to complex IV where they are passed from copper centre to haem a, haem a3 and second copper centre. Finally from complex IV, electrons are received by molecular oxygen, and water is formed.

Complex II is used when electrons enter via Flavin Adenine dinucleotide (FAD) and then go to Fe-S centres to Q. Succinate conversion to fumarate is a common source for these electrons.

It is important to note that Q can carry two electrons while cyt c can transfer only one, hence when one QH_2 is oxidised, two molecules of cyt c are reduced.

During flow of electrons across the complexes, protons are sent to intermitochondrial membrane space where a proton gradient is generated. Protons in the process of running through available spaces rotate ATP synthase enzyme embedded in the mitochondrial membrane, and ADP is phosphorylated. Thus oxidation of reducing equivalents is coupled to phosphorylation of ADP. This is termed oxidative phosphorylation. Molecules which interfere with ETC or coupling of oxidative phosphorylation are often fatal, for example, Cyanide and carbon monooxide. Genetic disorders involving ETC components lead to decreased ATP production and present with myopathy, fatigue and lactic acidosis (Fig. 3.1).

Sweet Biochemistry. DOI: https://doi.org/10.1016/B978-0-443-15348-8.00025-9

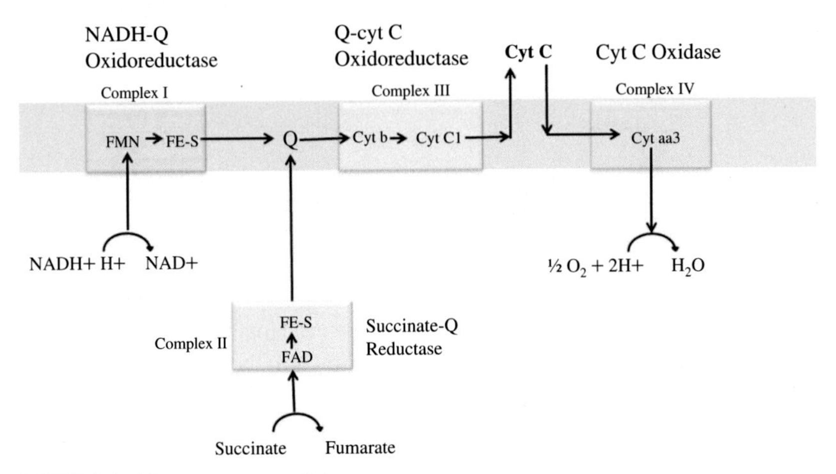

FIGURE 3.1 Electron transport chain.

Microview

To understand ETC, first you should know what is redox potential; otherwise the whole concept may become difficult. It will be like standing on a railway platform and watching the train just pass in front of your eyes. Without the information about arrival and destination cities, the journey becomes ambiguous.

So let's first familiarise with the redox potential. In simple words, it indicates the ease with which a molecule will accept electrons. This means that it is an inherent property of a molecule. More the positive redox potential, higher is the affinity of species towards electrons and easier to get reduced.

Imagine a country where tax is collected from poor farmers by landlords. From these landlords, tax goes to junior government officials who deposit it to their seniors. These senior officials then finally give all the tax collected from the country to the king. This is the simple sincere flow. Farmers cannot collect tax from the king, because they don't have the power. More powerful person can demand money from the weaker. Similarly, in ETC electrons flow towards the oxygen, the ultimate 'sink' of electrons having the most positive redox potential from molecules with lesser redox potential. But one thing which decreases from NADH to oxygen via complexes is potential energy. Hence, redox potential increases towards oxygen and potential energy falls. A redox potential of +1.14 volt exists between NADH/NAD + and H2O/1/2O2 passage.

Did you notice that the name of the chapter is electron transport chain. Why chain? Why not cycle like Krebs cycle? The large multisubunit enzyme complexes involved in the process are embedded in the inner mitochondrial membrane in a series for tight linking of numerous coupled redox reactions. Flow of reducing equivalents from electron donors to electron acceptor through this chain is known as electron transport. One NADH molecule brings two electrons which generate −52.4 kcal/mole free energy. This amount is much higher than that required to make one phosphoanhydride bond in ATP formation, which is −7.3 kcal/mole. So one NADH oxidation is able to generate a huge amount of free energy. Isn't it better to get this energy in one step why keep travelling via a long route via complexes? If this was done in one step in our body, the enormous amount of free energy would be released as heat only, and it was not possible to use it for our biological functions. When this one oxidation is split into many small reactions through carriers, almost half energy is conserved.

This free energy is used in producing terminal phosphoanhydride bond of ATP, justifying the name 'Oxidative phosphorylation'.

It is good to revise the brief anatomy of the mitochondria here. It is covered by two membranes separated by an intermembrane space. The central space inside the membranes is known as matrix. The permeability of these membranes is different, the inner membrane being less permeable allowing only small uncharged molecules to pass. Whole ETC system along with ATP synthase is embedded in the inner mitochondrial membrane.

Some more concepts should be clarified before the description of four complexes of ETC – the cytochromes. Cytochromes are metalloproteins, having haem as the prosthetic group performing the function of electron transport predominantly. Haem has iron atom in the centre which can shift between Fe^{2+} and Fe^{3+} states despite being connected to four nitrogens of the ring. When we see the diagram of ETC, there are Cytochrome a, b and c mentioned. These indicate the three groups of cytochromes classified on the basis of absorption maximum of the alpha band of a specific cytochrome. Haem in cytochrome b and c resembles the haem prosthetic group of haemoglobin and myoglobin where covalent linkage is formed between haem and apoprotein. However, the haem present in cytochrome a and a3 is different, as it is bound to protein part by hydrophobic interactions.

The special electron transporters: NADH and FADH2

Nicotinamide adenine dinucleotide (NAD) is a dinucleotide united through a phosphate group. Out of two nucleotides, one has Adenine base and second is Nicotinamide. NADH is generated during various oxidations in Glycolysis, fatty acid oxidation, etc. Therefore, it bring electrons from those pathways to finally ETC for ATP generation. NADH is oxidised by complex I of ETC, possessing NADH-CoQ reductase power. Ten protons are translocated by ETC when one NADH arrives – 4 from Complex I, 4 from Complex III and 2 from last Complex 4. ATP synthase, the enzyme phosphorylating ADP, requires four H + per phosphorylation of ADP. Therefore, after doing the mathematics, it is easy to calculate that one NADH will cause generation of 2.5 ATP.

FAD is structurally Adenine nucleotide joined with FMN via phosphate group. FADH2, the reduced form of FAD, is a potent oxidising agent, having more positive redox potential than NAD. The mechanism lies in the stability of aromatic ring of FAD which is greater than that in FADH2 state. Hence, when oxidation of FADH2 occurs, a large amount of energy is released to achieve stability. FAD produces 1.5 ATP (4 H+ translocated from complex III, 2 H+ from complex IV).

Ubiquinone or coenzyme Q

Coenzyme Q (CoQ) belongs to a coenzyme family ubiquitously present from bacteria to human, hence also referred to as Ubiquinone. Coenzyme Q10 is the most common in humans. These fat-soluble pigments are majorly found in the mitochondria. You must have read about vitamin K which is a phylloquinone. Still wondering what this molecule is? CoQ is 1,4 benzoquinone with a polyisoprenoid side chain attached at sixth carbon atom. Q represents quinone group in structure, and 10 number indicates number of isoprenyl chemical subunits in its tail. It is not tightly linked with a protein. Its importance in ETC can be guessed easily as it transports electrons from both complex I and II to complex III. CoQ can transform between oxidised, partially reduced and fully reduced forms. An antibiotic Antiycin A inhibits passage of electrons from QH2 to cytochrome C.

Complexes and electron flow

Inner mitochondrial membrane has large transmembrane proteins embedded in it which act as proton pumps. Three out of four such complexes, namely complex I, III and IV, can pump protons from the cytoplasmic side of the inner membrane to the intermembrane space. These pumps are connected by lipid-soluble ubiquinone and water-soluble cytochromes acting as electron carriers.

When electrons run through a complex, a current appears within the membrane (as current is generated by movement of charge). This electrical energy powers the active transport of protons. I hope you remember that active transport is mostly against concentration gradient. Then the crowding of protons creates a proton gradient which increases acidity here.

Now this proton gradient is like water stored behind a dam, which when allowed to run can rotate any turbine, and the kinetic energy of water flow is converted to electric energy.

Let's halt the proton gradient here and learn about complexes. We will return here again.

Complex I

You can call this large enzyme complex as NADH-CoQ reductase or Ubiquinone oxidoreductase. This protein has NADH dehydrogenase flavoprotein. There are two different electron carriers in this structure – FMN and iron-sulphur centre. FMN having Riboflavin vitamin is tightly bound prosthetic group of this enzyme. FMN is converted to FMNH2 by reduction with two reducing equivalents. FMNH2 becomes unstable and gets oxidised, passing electrons to Iron-sulfur(Fe-S) centres which further give them to Ubiquinone. In these non-haem Fe-S centres, Fe is covalently connected with cysteinyl sulfhydryl groups of protein. Fe is able to shift between Fe^{2+} and Fe^{3+} states here. Oxidation of iron is tied to reduction of Ubiquinone. Complex I is inhibited by a natural toxic plant product called Rotenone, a Barbiturate Amobarbital and an antibiotic Piericidin A.

Complex II

This complex enzyme is Succinate dehydrogenase, and this is also armed with a covalently bound FAD and Fe-S centres. Yes this is the same enzyme you read in previous chapter on TCA cycle. The donor of electrons to this complex is succinate. The oxidation of Succinate to Fumarate releases two electrons caught by FAD present bound to apoprotein (by Histidine amino acid residue) converting to the reduced form FADH2 transferred to Fe-S centre within this enzyme. This way also electrons reach Ubiquinone(CoQ) like the journey from complex I. However, a big difference to note here is that no protons are translocated by this electron transfer by complex II. Therefore the production of ATP is less than that in complex I path. Competitive Inhibitors of Succinate dehydrogenase are oxaloacetate and Malonate. Electron transfer from FADH2 to CoQ is blocked by Carboxin and Thenoyl trifluoroacetone.

Complex III

Other names for this multisubunit enzyme are Cytochrome C reductase, Cytochrome bc1 and Coenzyme Q: Cytochrome C oxidoreductase. Eleven subunits of this big protein can be divided into:

1. Three respiratory units (cytochrome b, cytochrome c1 and Rieske protein).
2. Two core proteins.
3. Six low-molecular-weight proteins.

This is the junction where electrons from oxidation of both NADH and FADH2 via Ubiquinone are received. As mentioned earlier, electrons from complex III are taken by water-soluble Cytochrome C to reach complex IV. Reaction catalysed here includes oxidation of CoQ along with reduction of cytochrome C. Meanwhile, four H+ are translocated to intermembrane space while intaking two protons from matrix. This contributes in forming a proton gradient.

Complex IV

Cytochrome C oxidase is the last complex in ETC which catalyses the final, rate-limiting step. Cytochrome a, Cytochrome a3 and two copper ions labelled CuA and CuB participate in redox reactions in this enzyme. Copper ions like iron ions can transfer electron. Actually one copper atom teams up with a haem (CuA with haem of cytochrome a and CuB with cytochrome a3). The electron moves from cytochrome C to CuA and then gives off to haem of cytochrome a. From here the electron is transported to haem of cytochrome a3-CuB binuclear centre. The oxygen molecule is also coordinated to iron atom and copper atom of this centre.

As oxidation of cytochrome C provides one electron, four rounds of this are required for combining four electrons, one dioxygen molecule and four protons brought from inner aqueous phase to form two water molecules. During this process, four protons are pumped in the intermembrane space.

So you notice that in this wonderful process designed by nature, protons are translocated during electron transfers. But you may think why it is like this? Reduction or addition of electron will bring charge on a molecule which can be neutralised if it can get one proton also. In that case $H+$ and electron addition will not add any net charge. Similarly if oxidation removes one hydrogen atom, it can break into $H+$ and electron which can travel separately, leaving neutral charge behind on molecule. In ETC, when electrons are received, the complex takes protons from matrix and while giving off electrons, protons are shifted to other side of the membrane, keeping charges on itself minimum.

I think now the whole route of ETC is clear to you. Still I will write a flow of electrons here for a nano summary (Fig. 3.2).

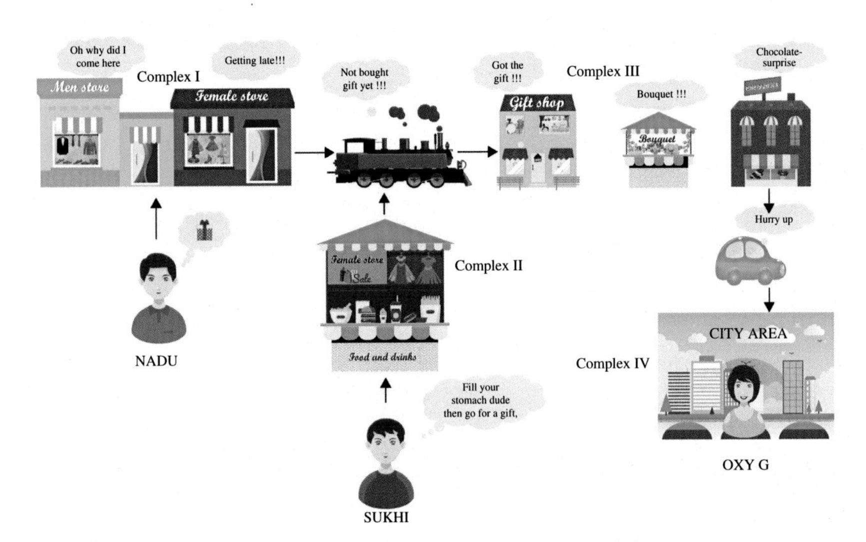

FIGURE 3.2 Pictoral representation of story for Electron transport chain.

NADH → FMN → Fe-S → CoQ → Cyt b → Fe-S → Cyt c1 → Cyt c → Cyt a-CuA → Cyt a3-CuB → O_2

Succinate → FAD → Fe-S → CoQ → Cyt b → Fe-S → Cyt c1 → Cyt c → Cyt a-CuA → Cyt a3-CuB → O_2

The story of electron transport chain

NADU wants to give gift to Oxy G, his girlfriend. As per reflex he enters men's store. He realises his mistake and proceeds to female store. He has not yet purchased the gift and the time of train boarding comes. He boards train towards destination. He again searches a gift in a female store and buys the gift. He adds a bouquet and chocholate from city shops. Then the NADU takes a car and reaches the city area 3 where Oxy G is waiting. Sukhi (NADU's friend) had suggested him to have Food and Drinks and then go to a female store and buy gifts.

What relates to electron transport chain in the story?

NADU → NADH
For MeN store → FMN
Female Store → Fe-S
Train → Coenzyme Q
City Bouquet shop → Cyt b
City Chocholate shop → Cyt c1
Car → Cyt C
City AreA3 → Cyt a a3
Oxy G → oxygen
Sukhi → Succinate
Food And Drinks → FAD

Chapter 4

Beta oxidation of fatty acids

Fatty acid oxidation is the mitochondrial aerobic process of breaking down a fatty acid into acetyl-CoA units. Fatty acid (FA) moves in this pathway as coenzyme A (CoA) derivatives using NAD and FAD.

The process starts with the uptake of fatty acid inside the cell. Then activation of FA is required to motivate it to enter the oxidation process. Acyl-CoA synthetase enzyme, the activator is located on the outer mitochondrial membrane surface and endoplasmic reticulum (ER). It carries out the ATP-dependent thioesterification of Fatty acids with Coenzyme A.

Long-chain acyl-CoA enters the mitochondria bound to Carnitine – a quaternary ammonium compound. The entry is itself the rate-limiting step of beta oxidation. Inside the mitochondria beta oxidation of fatty acids takes place in which two carbon atoms are removed in the form of acetyl-CoA from acyl-CoA from carboxyl terminal. Bond is broken between second carbon/beta carbon and third carbon/gamma carbon, hence the name beta oxidation is given.

First, FAD-dependent dehydrogenation is done by Acyl-CoA dehydrogenase which results in double bond formation between C2 and C3. $FADH_2$ is generated in this reaction. In the next step, water is added by Enoyl-CoA hydratase on double bond forming 3-hydroxyacyl-CoA. Second, NAD + -associated dehydrogenation by 3-hydroxyacyl-CoA dehydrogenase converts hydroxy group on C3 to keto group yielding 3-ketoacyl-CoA.

Thiolase enzyme cleaves the bond between C2 and C3 releasing acetyl-CoA and acyl-CoA which is two carbon atoms shorter than the starting molecule. This new acyl-CoA enters the same pathway again.

In this way, the Acetyl-CoA is sequentially removed from acyl-CoA until 2 acetyl-CoA are left in the end. Odd-chain fatty acids leave acetyl-CoA and Propionic acid on completion. Acetyl-CoA is directed to citric acid cycle for further oxidation. One cycle of beta oxidation releases one FADH2 and one NADH + which causes four high-energy phosphate bond synthesis in ETC along with one acetyl-CoA and acyl-CoA that is two carbon atoms shorter. Beta oxidation of fatty acids thus provides a large number of ATP (one palmitic acid provides approximately 106 mol ATP per molecule of palmitic acid). One fatty acid molecule faces multiple rounds of beta oxidation till full chain is broken into acetyl-CoA (in odd-chain FA, propionyl-CoA is left along with 2 C Acetyl-CoA) (Fig. 4.1).

Sweet Biochemistry. DOI: https://doi.org/10.1016/B978-0-443-15348-8.00021-1

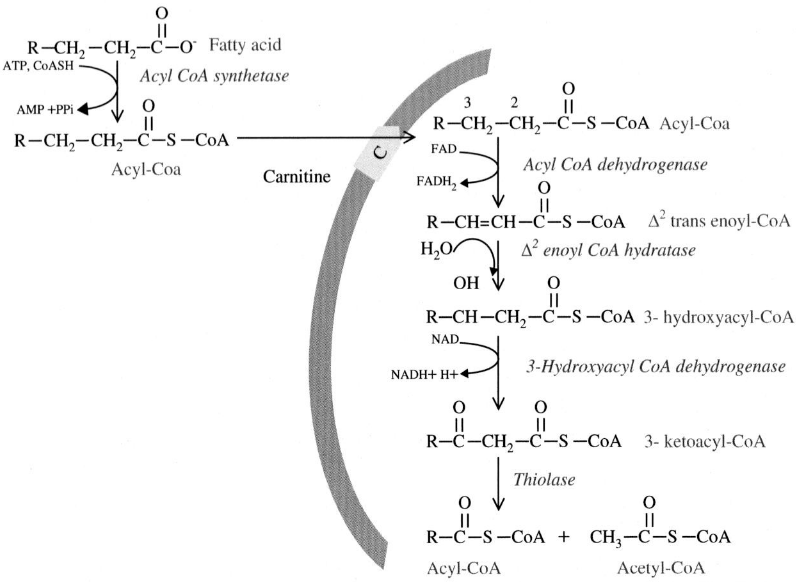

FIGURE 4.1 Beta oxidation of fatty acids. Yellow block with pointed end represents Carnitine.

Importance of beta oxidation of fatty acids

Fatty acids are among the most important biofuels. Therefore, the oxidation of fatty acids holds a crucial position in metabolism. When glucose levels fall between the meals, during exercise, or increased requirement states, fatty acid is released from fat cells on getting signals from Glucagon and epinephrine. Skeletal muscles, heart, and kidneys consume large amount of fatty acids during this time (Glycogen breakdown and gluconeogenesis are not able to maintain energy requirement). You might have heard many times that first fuel used is glycogen, then fatty acids and finally proteins are channelled in energy production. Thus fatty acids protect proteins from catabolism for gluconeogenesis.

In patients with Diabetes Mellitus, when glucose is not able to enter the cells, beta oxidation of fatty acid is the rescuer. Acetyl-CoA released enters ketone bodies synthesis which acts as energy substrate for many tissues including the brain. By providing substrates for gluconeogenesis, beta oxidation of fatty acids prevents hypoglycemia. This becomes evident in defects of beta oxidation pathway, for example, in carnitine deficiency, CPT enzyme defect or hypoglycin poison.

Fatty acids can be oxidised by alpha, beta and omega-oxidation. Alpha-oxidation takes place in peroxisome, beta oxidation in mitochondria and peroxisome, and omega-oxidation occurs in Endoplasmic reticulum. Peroxisome uniquely deals with beta oxidation of very-long-chain fatty acids, which are

24–26 carbon chains long. Sole purpose of Fatty acid oxidation is not just providing energy, rather other functions include removal of large, insoluble xenobiotic compounds and lipid-based cell components. The product of pathway Acetyl-CoA enters different fate depending on the organ. In muscles, acetyl-CoA generated enters TCA for energy production while in the liver, it is used for ketone bodies synthesis during starvation or fasting when glycogen stores have been consumed.

Carnitine: prominent molecule of beta oxidation

Carnitine word is rooted to *Carnis* which means flesh. The liver and kidney synthesise this molecule from two essential amino acids Lysine and Methionine. Carnitine is then ferried to skeletal and cardiac muscles as these tissues cannot make it. Its chemical structure is L-β-hydroxy-γ trimethylammonium butyrate. Only L form is found in animals naturally (Fig. 4.2).

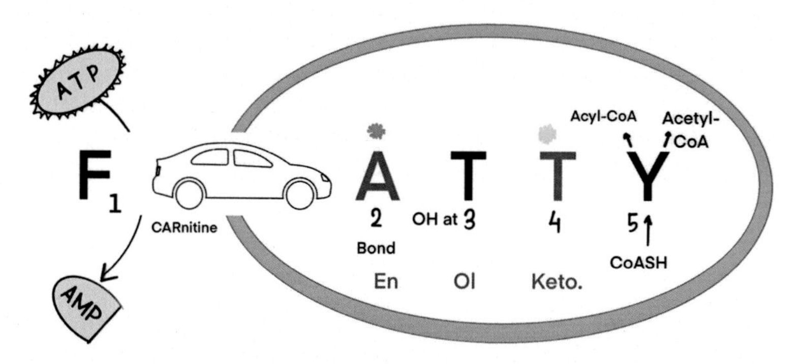

Beta oxidation of fatty acids includes five principal reactions. Write FATTY and numerate them.
Reaction 1 Is activation of fatty acid using 1ATP releasing AMP. Carnitine, like a car transports long chain fatty acid inside the mitochondria.
Reaction 2 is a 2 bond (EN) formation: FAD-dependent dehydrogenation reaction.
Reaction 3 is OH group (OL) formation on 3rd C by hydration.
Reaction 4 is =O (keto) group arrival on 3rd C: NAD+ linked dehydrogenation.
Reaction 5 is breaking bond between 2C & 3C with release of acetyl-CoA and Acyl-CoA.

Remember EN→OL→KETO.

FIGURE 4.2 Illustrated mnemonic for beta oxidation of fatty acids. Important reactions are summarised in this figure.

The role of Carnitine becomes paramount in oxidation of long-chain FA as the transport of LCFA is difficult as compared with short-chain and medium-chain FA. L-carnitine acetyltransferases are the enzymes which perform the reversible transfer of acyl group between L-carnitine and CoA. This shuffling allows transport of acyl-CoA esters from the cytoplasm through the impermeable mitochondrial membrane to the matrix side. The set of enzymes involved in this molecule transport is called Carnitine palmitoyltransferase system. These enzymes include Carnitine palmitoyltransferase (CPT) 1 and 2, carnitine–acylcarnitine–translocase (CACT) and acyl-CoA synthetase (counted by some authors).

It is good to highlight some details about L-carnitine acetyltransferases which are promising candidates for new drugs targeted for inflammatory pathways and cancer. Transmembrane CPT has two isoforms. CPT1 is present on outer mitochondrial membrane (OMM), and CPT2 is on the inner side of Mitochondrial membrane. CPT1 is further divided into three isoforms depending on the tissues: L-CPT-1, CPT1A in the liver, M-CPT-1, CPT1B in the muscle and B-CPT-1, CPT1C in the brain. Location of CPT1 and CPT2 is as per their role. CPT1 on the OMM transesterifies acyl group from acyl-CoA to acyl-Carnitine and releases CoA in the intermembrane space deciding the rate of whole beta oxidation. I hope that you remember that there is a intermembranous space between two membranes of the mitochondria. Now wandering Acyl-Carnitine is picked by CACT and shifted to matrix. CACT can send free Carnitine back to the intermembrane space. On the matrix side of IMM, acyl-Carnitine is converted by CPT 2 into acyl-CoA for oxidation.

CPT system thus plays a key role in the energy production pathway of body. Energy production being a chief requirement of tumour cells, CPT system is now researched for various drug targets (Fig. 4.3).

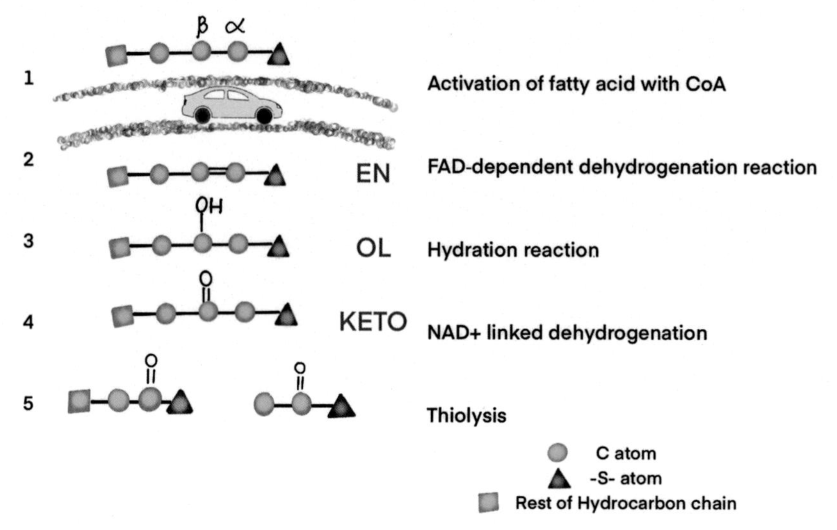

FIGURE 4.3 Another simple way to understand the basics of the reactions of beta oxidation. The yellow car inside mitochondrial space represents Carnitine, that transports long-chain fatty acids.

Chapter 5

Fatty acid biosynthesis

Traditional recap

Fatty acids are synthesised in the cytoplasm with participation of ATP, NADPH, Biotin, HCO_{3-} and Mn++. Major sites of this pathway are the liver, kidney, brain, lung, mammary gland and adipose tissue where HMP pathway can provide ample amount of NADPH. Acetyl-CoA is the two-Carbon building block molecule which is added to Malonyl CoA (formed by carboxylation of Acetyl-CoA). Enzyme catalysing this reaction is the regulatory enzyme of fatty acid synthesis – acetyl-CoA carboxylase. This process is cytoplasmic while beta oxidation takes place in the mitochondria. Main substrate for biosynthesis is glucose; that is why high-carbohydrate diet promotes lipogenesis. Fatty acids synthesis is a very crucial pathway in view of alarming increase in obesity and diabetes mellitus. Fatty acids are precursors of eicosanoids, complex lipids, membrane lipids and second messengers of signal transduction regulating cell functions.

First step is transfer of acetyl-CoA on Cysteine-SH group by Acetyl transacylase and malonyl-CoA on 4′-phosphopantetheine of ACP by malonyl transacylase. These two sites belong to different units of dimer enzyme complex called Fatty acid synthase (FAS). This complex enzyme has seven enzyme activities on both monomer and hands over substrate from one component enzyme to the other. This improves the efficiency of reaction.

In the second step, acetyl group joins malonyl residue by 3-ketoacyl synthase, and CO_2 is released yielding 3-ketoacyl enzyme. One site is now free. 3-Ketoacyl enzyme is reduced using NADPH and enzyme 3-keto acyl reductase forming 3-hydroxyacyl enzyme. Fourth step is dehydration catalysed by hydratase. This step creates double bond between C2 and C3. This bond is reduced by enoyl reductase enzyme. Second reduction also needs NADPH.

Finally, acyl-S-Enzyme is formed. Now this acyl residue is shifted to empty Cysteine-SH group and a new malonyl residue arrives at –SH of 4′-Phosphopantetheine. The aforementioned five reactions repeat until the desired length is achieved. Thioesterase releases the completed fatty acid by hydration.

Linoleic and Linolenic fatty acids are called essential fatty acids as they cannot be synthesised by humans (Fig. 5.1).

Sweet Biochemistry. DOI: https://doi.org/10.1016/B978-0-443-15348-8.00031-4

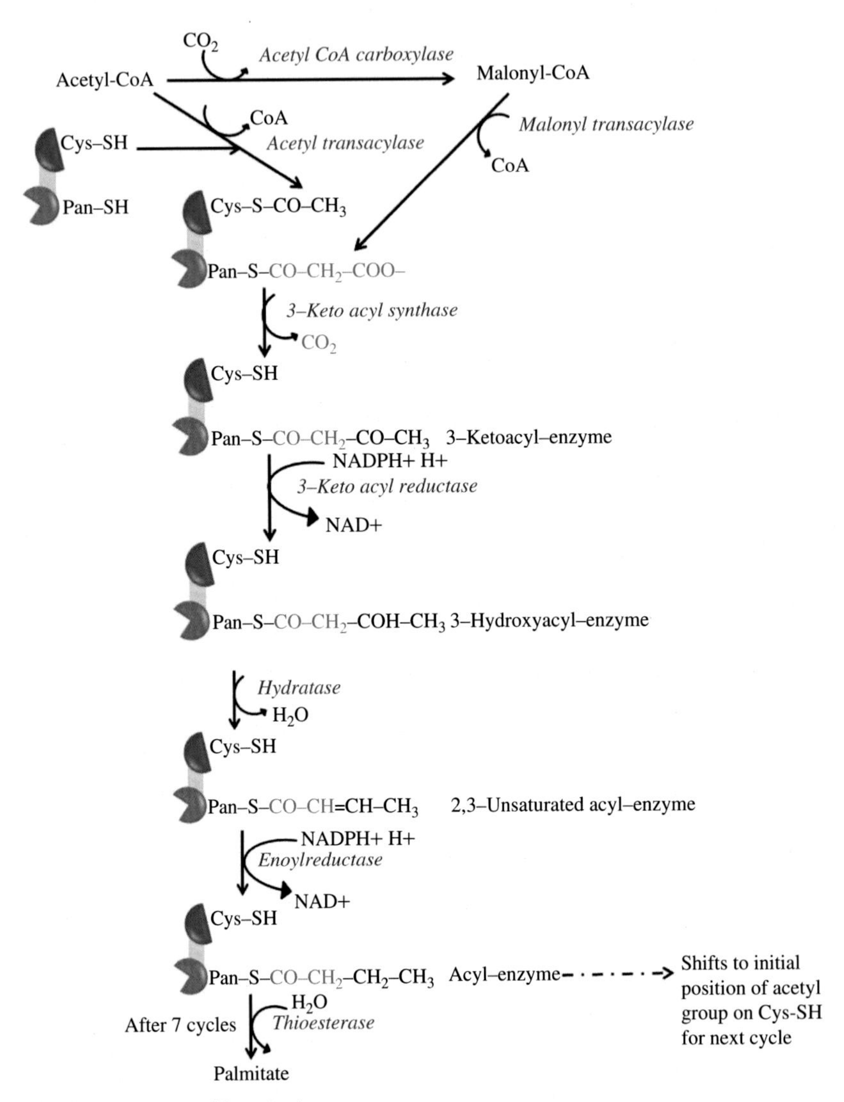

FIGURE 5.1 Fatty acid synthesis.

Microview

Animals store excess calories in the form of fatty acids. Oxidation of one palmitic acid molecule provides 129 ATP while one glucose yields 30–32 ATPs only. Therefore, fatty acids are a dense source of energy. When lot of energy accumulates, cell makes fatty acids, but how would a cell know that energy is in plenty? Acetyl-CoA, AMP and presence of citrate indicate abundance of energy. This energy is packed with the help of reducing equivalents (NADPH) which are derived from pathways such as Pentose Phosphate Pathway (PPP). PPP runs in the cytoplasm, therefore it would be convenient if acetyl-CoA can come here. But you remember from TCA chapter that acetyl-CoA is produced in the cytoplasm.

Then it came out? How? There are two paths to choose. In one, citrate (acetyl-CoA combined with oxaloacetate) comes out of the mitochondria by citrate shuttle, and in the second way, acetyl-CoA combines to carnitine and escapes to the cytoplasm. Carnitine is the same molecule which helps long-chain fatty acids to enter the mitochondria for oxidation. (Carnitine is up to something, participating in beta oxidation of fatty acids and here in fatty acid synthesis also! Have you heard of god who creates world with one hand and destroys with other?)

The cells want to join carbon building blocks, but molecules don't combine that easily. Molecules need to be activated by adding energy for producing new bonds in the presence of some efficient enzymes. Acetyl-CoA carboxylase and FAS are two such multifunctional polypeptides. We need to familiarise with both, then the journey of fatty acid synthesis will become smooth.

Acetyl-CoA carboxylase

Biotin acts as a prosthetic companion of this enzyme linked in amide bond to the e-amino group of a lysyl residue. It is present in an inactive protomer state with molecular weight of 400,000. Activation is done by polymerisation and reaching weight of 4–8 million. Then activities performed are biotin carboxylase, biotin carboxyl carrier protein and transcarboxylase. Acetyl-CoA carboxylase is under short-term allosteric regulation with citrate, isocitrate favouring activation while the product of fatty acid synthesis pathway, long-chain acyl-CoA derivatives causing inhibition by binding to allosteric site. Glucagon and insulin cause phosphorylation and dephosphorylation of Acetyl-CoA carboxylase, respectively, which consecutively lead to inhibition and activation of enzyme.

It is well known that high carbohydrates promote fatty acid synthesis. It is by long-term regulation of acetyl-CoA carboxylase along with fat-free diet, deficiency of choline and vitamin B12. High-fat content in diet, fasting, PUFA intake and deficiency of Biotin on the other hand lower activity.

Fatty acid synthase

It is the largest multifunctional protein, analogous to a workstation of laboratory where one substrate is passed from one stage to other on closely placed machines. In bacteria, the component subunits of FAS are found independently. However, in higher organisms, it united as a complex so that one gene can express all enzymes in equal amounts, substrate can interact with ease, resulting in greater efficiency. FAS possesses seven different catalytic actions performed by three different globular domains. These are

Condensing domain, Reduction domain and Releasing domain. The enzymes present in each one are:

1. Condensing domain – Acetyltransferase/Acetyl transacylase (AT) and Malonyltransferase/Malonyl transacylase (MT), Beta ketoacyl synthase (KS).
2. Reduction domain – Dehydratase (DH), Enoyl reductase (ER), beta-ketoacylreductase (KR) and Acyl carrier protein (ACP).
3. Releasing domain – Thioesterase (TE).

From the names of enzymes and domains, you should guess their actions. Organisation of FAS has another level of homodimer form where each unit is joined head to tail and tail to head. (Don't get confused, the FAS was trying to copy double-helix anti-parallel DNA structure, LOL). Monomer cannot do action of beta-ketoacyl synthase, which is possible in dimer form only. More surprising is the fact that synthesis of fatty acid can occur from any head of monomer unit.

I think we want to delve deeper. I got a yes from you!!

Fatty acids are made inside cells by repetition of decarboxylative claisen condensation reaction which joins acetyl-CoA and malonyl-CoA. The product has beta-keto group which needs reduction to form a saturated hydrocarbon chain with C-C bond. This is achieved by many rounds of actions of ketoreductase, dehydratase and enoyl reductase.

First step catalysed by acetyl-CoA carboxylase is the rate-limiting step.

This malonyl-CoA is a carboxylated acetyl-CoA only, for whom carbon is not from carbon dioxide but bicarbonate, in the presence of biotin, ATP and acetyl-CoA carboxylase enzyme as mentioned earlier. The rate-limiting step is facilitated by citrate by promoting conversion of inactive Acetyl-CoA carboxylase enzyme to active. Biotin is not wandering freely, rather it is attached to Biotin carboxyl carrier protein part, which accepts COO- from HCO3- by Biotin carboxylase action, consuming energy from ATP. Then by Transcarboxylase action, COO- is transferred to acetyl-CoA, forming Malonyl-CoA.

Now both acetyl- CoA and Malonyl-CoA are ready, but they need to be loaded on workstation FAS at Acyl carrier site (ACP) by letting CoA go. AT and MT enzymes attach Acetyl-CoA on cysteinyl-SH group and Malonyl-CoA on 4′-phosphopantethiene group of ACP. Both these attachments are not present on one unit of FAS but different units.

Next step catalysed by 3- Ketoacyl synthase is a decarboxylative claisen condensation which actually releases CO_2 and then combines carbon units. Decarboxylation acts as the crucial driving force for the reaction. Acetyl-CoA (2C) is Claisen's electrophile while Malonyl-CoA (3C) is Claisen's nucleophile. Claisen condensation is the formation of beta-diketone or beta-ketoester by joining one ester and one carbonyl compound in the presence of strong base. By this reaction, thioesters inside the cell are condensed. Sulphur atom of thioester from CoA is approximately 100 times more acidic than if Hydrogen would have been here. The product of condensation is acetoacetyl-ACP.

Now the main work is done, the formation of a longer chain but fatty acid is not like this. To make it saturated, some more steps are required. First reduction of carbonyl group by stereospecific Ketoacyl reductase using first NADPH adds hydrogen atom forming hydroxyl group, yielding D(-)-β-hydroxylbutyryl-ACP or β-hydroxy fattyacyl-ACP. In the next step, Dehydratase extracts a water molecule, resulting in α,β-unsaturated acyl-ACP. This unsaturation can be easily saturated by one reduction, so here it is performed by Enoyl-reductase generating butyryl-ACP (second NADPH used here).

The chain formed up to this stage is shifted to -SH group of condensing domain on second monomer. New Malonyl-CoA binds to 4′-phosphopantethiene group of ACP which is condensed to this chain, and again the cycle of condensation, reduction, dehydration and second reduction continues followed by shifting to other monomer. This repetition ends when desired length of fatty acid is achieved. Here comes the action of Thioesterase which cleaves palmitate (for example) from 4′-phosphopantethiene group of ACP. In the mammary gland, the thioesterase acts on fatty acids with length of C8, C10 or C12.

Fatty acid synthesis can be compared with depositing money (carbon atoms) in a bank (Fig. 5.2).When desired amount has been gathered, the money can be taken out. Two enzyme sisters Acy Tracy and Mala Tracy deposit their cash (carbon atoms) in a bank. Acy tracy deposits in CYS bank. Mala tracy deposits her cash on the same PAN card in a different branch. Then both wanted to shift their money in joint account so Cash (KAS) officer transferred Acy's carbon on the same PAN card but kept one carbon as bribe. On the PAN card the cash followed the sequence of keto → ol → en (Fig. 5.3).

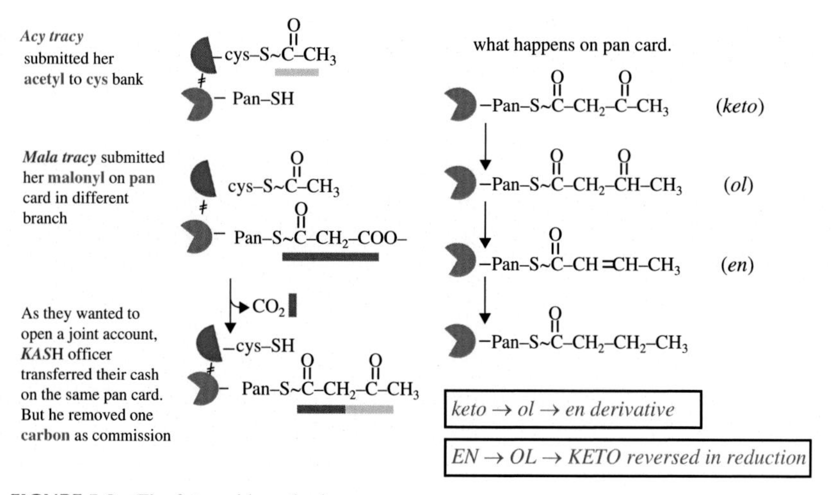

FIGURE 5.2 The fatty acid synthesis story.

What relates to fatty acid biosynthesis in story?

	Ketoacyl synthase
	Acyl carrier protein (ACP)
Acy tracy	Acetyl transacylase
cys bank	Cysteine —SH group of Ketoacyl synthase
Mala tracy	Malonyl transacylase
pan card	—SH on the 4′-phosphopantetheine of ACP
KASH	3-ketoacyl synthase

FIGURE 5.3 Links of story with fatty acid synthesis.

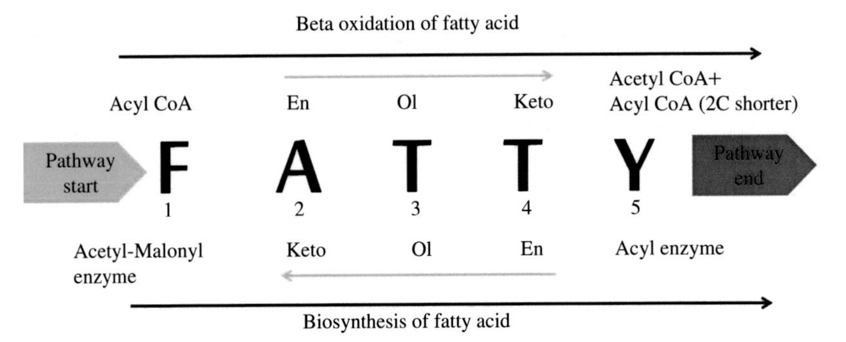

FIGURE 5.4 Beta oxidation and synthesis of fatty acids.

Relation of beta oxidation and synthesis of fatty acids synthesis

Fatty acid synthesis and oxidation are not exactly the opposite process, and their location is also different. Synthesis is cytosolic, and oxidation is mitochondrial. However, some note that fatty acid synthesis and beta oxidation both have five important steps. These two processes run in opposite direction as shown earlier (Fig. 5.4).

In beta oxidation, fatty acid is activated by CoASH, and then en → ol → keto derivatives are formed. Then acetyl-CoA is removed.

Opposite happens in the synthesis where acetyl-CoA is added to growing acyl group, and keto → ol → en derivatives are formed which are finally reduced.

Chapter 6

Cholesterol structure

Out of all the bad molecules, cholesterol stands out probably as the worst villain. We are trying to eat everything free of cholesterol, be it your burger, pizza or milk. As if it is some kind of a weak poison. Actually the situation is not like this. Just like the DNA, RNA or the enzymes, cholesterol is a wonderful biomolecule, very much required for the optimum cell functioning. Even if the dietary cholesterol in diet is reduced to zero, the plasma concentration will not drop to the same. Why? Because our body needs it. Every human cell can manufacture the cholesterol required for it. Therefore, we are not the Spiny lobsters or *Caenorhabditis elegans* that need cholesterol from diet.

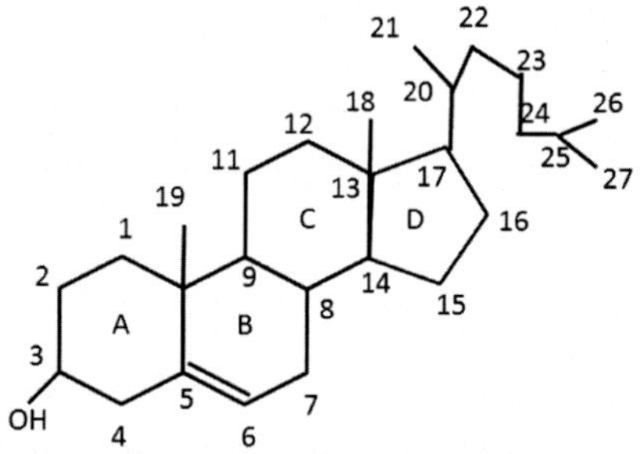

FIGURE 6.1 Cholesterol structure.

Incorporated in the cell membranes, cholesterol performs some of the most amazing functions, making the life possible. The long list of biological functions is a result of unique structure of cholesterol. You can credit it as a gift to the animal cell membrane, differentiating it from plant cell membrane which needs to be covered with a cell wall.

The name Cholesterol indicates both the important structural entities alcohol and steroid. Chemically, it is 3-hydroxy-5,6-cholestene, with a flexible isooctyl hydrocarbon tail, central sterol nucleus and a hydroxyl group. The steroid nucleus is bulky, rigid, planar tetracyclic ring structure also known as cyclopentano-perhydro-phenanthrene. The rings are labelled as ring A, B, C and D, first three being six-membered and last one is five-membered. If you see the numbering of cholesterol in Fig. 6.1, there are two

Sweet Biochemistry. DOI: https://doi.org/10.1016/B978-0-443-15348-8.00023-5

methyl groups attached – one at C18 and C19. These side branches are attached at right angle to the rings above the plane in beta position. Methyl groups increase the interactions with other hydrophobic lipids.

There is a long eight-carbon hydrocarbon tail at C17. Due to the hydrophobic nature of this tail, it intercalates in the nonpolar core of bilayer membranes up to the depth of C9–C10 of fatty acid component of phospholipid. A double bond is found between C5 and C6 in ring B, and its presence ensures that the tetracycle becomes planar. A hydroxyl group at C3 is responsible for some polarity in this hydrophobic molecule which facilitates the position of cholesterol in phospholipid bilayer. This is also the site for esterification with fatty acid or forming hydrogen bonds with neighbouring carbonyl oxygen present in phospholipid and sphingolipid head.

Role of cholesterol

Cholesterol, due to its amphiphilic nature, modulates fluidity and permeability of the membrane. Fluidity is the movement of molecules in the plane of phospholipid bilayer, bestowing the flexibility. (Allowing our cells to change shape and move. Think for a moment what will happen if all the cell membranes become stiff?) The factor affecting this is the lipid composition including that of phospholipids and cholesterol in the membrane. Cholesterol prevents the membrane stiffening at low temperatures by disturbing the tight packing of phospholipids. At high temperature, there is a risk of excessive fluidity. This is reduced by cholesterol due to its rigid planar structure.

Another factor involved here is that interaction between cholesterol and heads of phospholipids immobilises the proximal areas of hydrocarbon chain. This interaction also prohibits the passage of polar molecules across the membrane. Content of cholesterol varies with the location of the membrane. Membranes with the primary function of forming compartments such as cell organelles usually have low content of cholesterol. On the other hand, high cholesterol content helps in the formation of hydrophobic channels for passage of small, hydrophobic molecules while transport of these is not allowed across the membrane due to rigid structure.

Lipid rafts and cholesterol

Lipid rafts are Cholesterol and sphingolipid-rich, liquid-ordered microdomains. Traditionally it is defined as a detergent-resistant membrane fraction. Saturated hydrocarbon chains in sphingolipid and phospholipids lead to tighter packing as compared with the other parts of the membrane (You can compare them to boats floating on lakes). Rest of the bilayer predominantly has unsaturated fatty acids in phospholipids which favours loose packing due to the presence of double bond.

These rafts have integrations with some big proteins by covalent bonds to glycolipids and Glycosylphosphatidyl-inositol. These rafts are dynamic which means that proteins and lipids are free to move in and out from here. The proteins present in the raft can act as a receptor or a channel, thereby getting involved in numerous cellular functions. Important role of lipid rafts has been discovered in signal transduction especially in hematopoietic cells. Immune receptors such as IgE receptor, T-cell and B-cell receptors are found to move to the lipid rafts upon cross-linking. Sorting of proteins inside the cell, regulating the distribution of receptors in the membrane and affecting cell surface proteolysis are other pathways where lipid rafts participate. Endocytosis mediated via caveolae and virus binding is also associated to these domains.

Apart from forming lipid rafts, the presence of cholesterol – the ubiquitous molecule of membrane, is noted in abundance in the brain and myelin sheath. Cholesterol works in the formation of synapses and therefore is crucial in memory and learning ability. Some other well-known functions of cholesterol are being a biofuel, Thermo regulation, protection of internal organs, acting as precursor of steroids and cortisone-like hormones.

I imagined the cholesterol structure as a honeycomb house of a queen bee as the six-membered rings of cholesterol are hexagonal. So look at a unique cholesterol house of Queen Bee.

What relates to cholesterol in honeycomb house?

bedroom: three cyclohexane rings,
one kitchen: one cyclopentane ring,
double bed: double bond between C5 and C6,
OH group at entrance: OH group at C3,
on the roof, two methyl vents: methyl groups at C10 and C13,
Antena: eight-carbon side chains on C17.

This picture helps to remember cholesterol structure very easily if you visit each corner of the house vigilantly. Try to sing (Fig. 6.2).

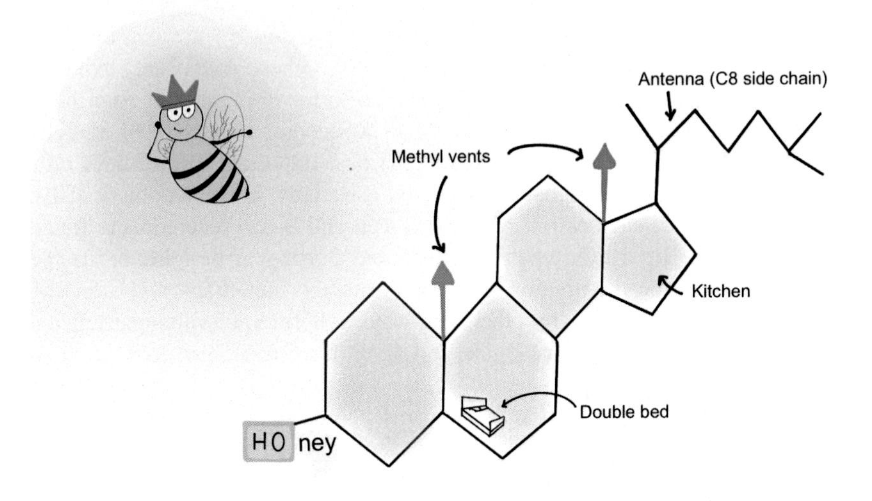

3 Bedroom and 1 kitchen set
Between C5 & C6, there is a double bed (bond)
OH group at the entrance
On the roof, 2 methyl vents
8 carbon antenna for TV
Is adjusted by the Queen Bee

FIGURE 6.2 Cholesterol structure visual mnemonic and a rhyme of queen bee.

Chapter 7

Cholesterol synthesis

Traditional recap

Cholesterol synthesis is an expensive process for cell in terms of energy. These pathways take place in the cytoplasm. Liver and intestines are major contributors of endogenous production. Hypercholesterolaemia, typically elevation of LDL and Oxidised-LDL have been implicated in cardiovascular and cerebrovascular events. Hence, it is indispensable to learn about the pathway of cholesterol synthesis.

Acetyl-CoA units are joined to form a 30-carbon compound, and then three carbons are removed to produce cholesterol with 27-carbon atoms.

Steps of cholesterol synthesis can be divided into:

1. Mevalonate synthesis.
2. Isopentenyl phosphate synthesis.
3. Squalene formation.
4. Lanosterol synthesis.
5. Cholesterol formation.

Two acetyl-CoA combine to form acetoacetyl-CoA releasing CoA-SH in the presence of Thiolase enzyme. Third Acetyl-CoA also condenses to form 3-hydroxy-3-methylglutaryl-CoA (HMG-CoA) catalysed by HMG-CoA synthase. These enzymes are different from the enzymes used for ketone bodies synthesis in the mitochondria. HMG-CoA is reduced by HMG-CoA reductase using NADPH to Mevalonate. This enzyme is the regulatory enzyme of pathway, inhibited by statins the widely used lipid-lowering drugs.

Mevalonate is phosphorylated by three kinases sequentially using three ATP and then decarboxylated to form isopentenyl diphosphate.

Isopentenyl diphosphate (5C) isomerises to 3,3 Dimethylallyl diphosphate (5C) by shifting a double bond, and then the condensation with isopentenyl diphosphate forms Geranyl diphosphate (10C). Another isopentenyl diphosphate molecule joins to form 15C compound – Farnesyl diphosphate. Two such 15C molecules fuse to form 30C Squalene.

Squalene is oxidised to Squalene 2,3-epoxide by squalene epoxidase. During cyclisation to Lanosterol, methyl group shifts from C14 to C13 and from C8 to C14.

The methyl groups on C14 and C4 are removed to form 14-desmethyl lanosterol and then Zymosterol. The double bond at C8–C9 is subsequently

Sweet Biochemistry. DOI: https://doi.org/10.1016/B978-0-443-15348-8.00026-0

shifted to C5–C6 in two steps, forming Desmosterol. Final step is the reduction of double bond of side chain yielding cholesterol molecule (Figs. 7.1–7.3).

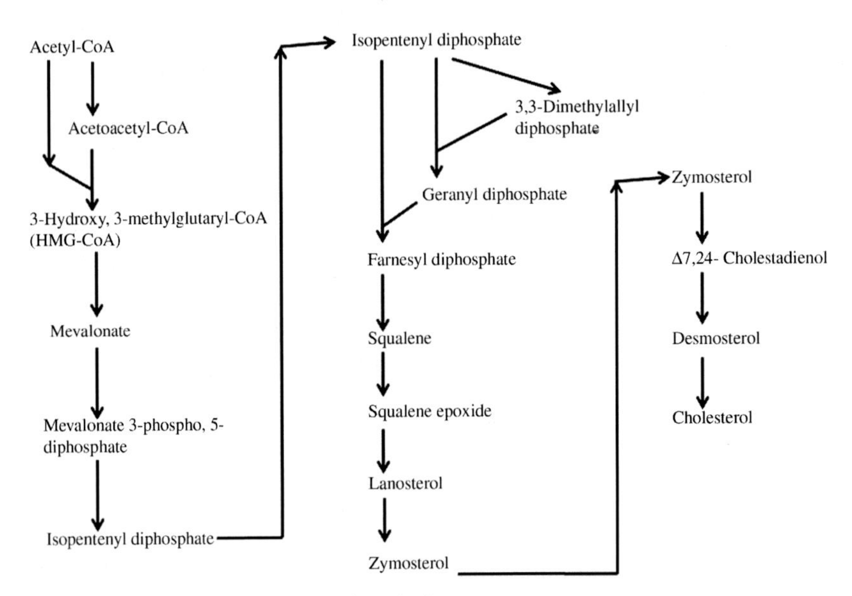

FIGURE 7.1 The basics of cholesterol synthesis.

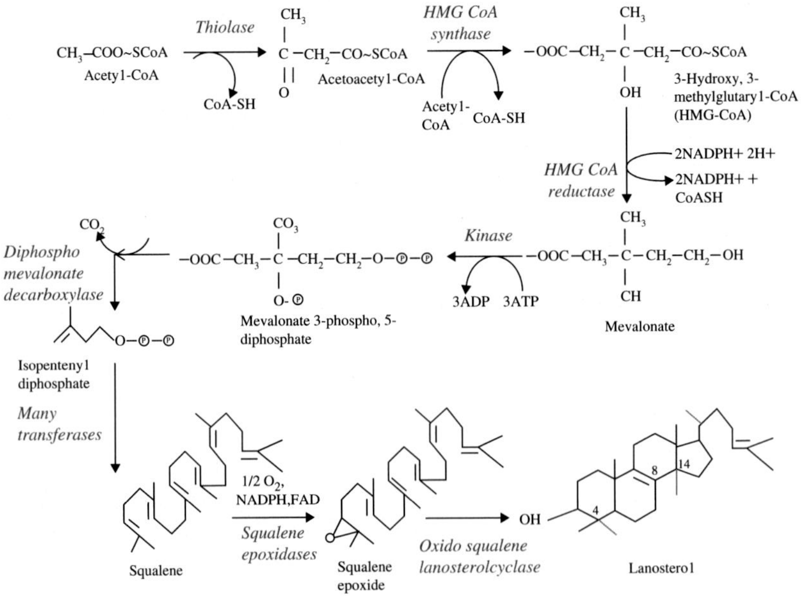

FIGURE 7.2 Cholesterol synthesis steps, part 1.

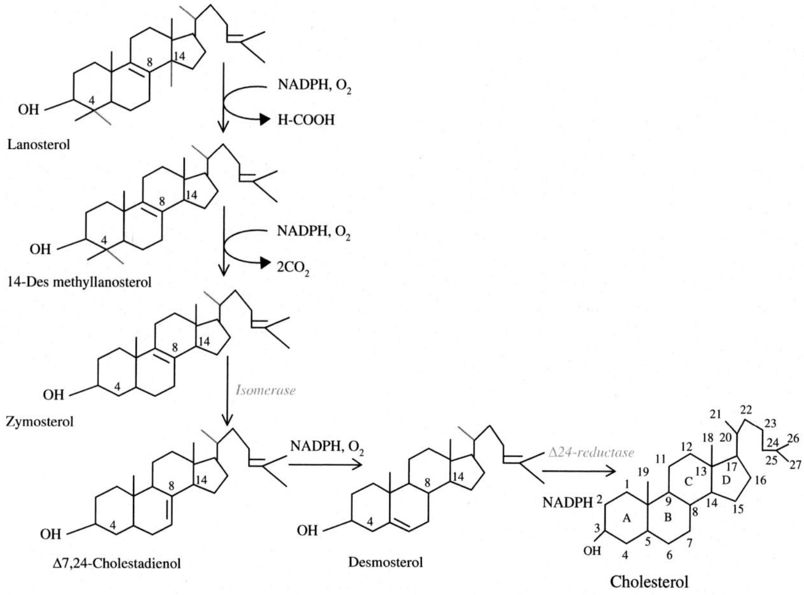

FIGURE 7.3 Cholesterol synthesis steps, part 2.

Therefore, the long sequence is $C2 + C2 = C4$, $C4 + C2 = C6$, $C6 - C = C5$, $C5 + C5 = C10$, $C10 + C5 = C15$, $C15 + C15 = C30$, $C30 - 3C = C27$.

Microview

Cholesterol synthesis starts with acetyl-CoA. Acetyl-CoA has a thioester linkage which on hydrolysis releases lot of free energy. Basically acetyl-CoA is an energetic molecule which can be used for many biosynthesis processes. In fed state, acetyl-CoA increases in the nucleus and cytoplasm, subsequently can form fatty acids or cholesterol by getting diverted away from the mitochondria. During fasting in the mitochondria, TCA will absorb acetyl-CoA, or it can form ketones.

First enzyme thiolase

Thiolase carries Claisen condensation of two acetyl-CoA molecules in three steps. You will see this reaction in many biosynthetic pathways. Thiolase acetylates its special cysteine residue at an active site by acetyl-CoA. In the second step, the enzyme attaches second acetyl-CoA and binds, and enolisation takes place. Third step involves Claisen's condensation between two acetyl-CoA still hanging at the cysteine residue. (Imagine how an enzyme will glue two substrates held in its amino acid residues.) Acetoacetyl-CoA is formed here.

3-Hydroxy-3-methylglutaryl-CoA synthase

HMG-CoA synthase exists as two main isozymes: a mitochondrial ketogenic form and a cytosolic cholesterogenic form. This enzyme adds water to join acetoacetyl-CoA with one more acetyl-CoA, removing 1 CoA-SH meanwhile. Like the previous enzyme, this reaction is basically combination of similar three steps: acetylation of enzyme, condensation and hydrolysis. A vital catalytic site cysteine residue of HMG-CoA synthase works as a nucleophile during acetylation of enzyme. Then Claisen condensation takes place between acetoacetyl-CoA and acetyl-CoA. Finally, the thioester linkage is broken and 3-hydroxy-3-methylglutaryl-CoA (HMG-CoA) is produced.

3-Hydroxy-3-methylglutaryl-CoA reductase

The rate limiting of cholesterol is catalysed by HMG-CoA reductase which converts cytosolic HMG-CoA into Mevalonate. Being the target of statins – most important lipid-lowering drugs, the regulation of HMG-CoA reductase becomes important. Embedded in the membrane of the Endoplasmic reticulum, the single polypeptide enzyme has two domains: a sterol-sensing domain which is the membrane-bound component and a catalytic domain which is soluble. Names are clearly indicating the functions. Sterol sensing is important probably because most of the times this enzyme is under competitive inhibition, and you know what are the inhibitors – cholesterol and oxidised variants of cholesterol derived from LDL which was taken inside the cell. Thus it is quite clear that endogenous production of cholesterol is fine-tuned with the dietary intake. Our body can raise the production several hundred fold in case of low dietary intake. Active site of enzyme is present at the junction of a homodimer between monomer-binding NADPH and HMG-CoA.

The committed step of mevalonate synthesis takes place in three steps. (Glad to see such uniformity of numbers among enzymes of a pathway!) Among these, two are reductions, and mevaldyl-CoA and mevaldehyde are the intermediates. Simplified reaction steps are mentioned as follows:

1. $\mathrm{HMG-CoA + NADPH + H+ \rightarrow Mevaldyl-CoA + NADP+}$
2. $\mathrm{Mevaldyl-CoA \rightarrow Mevaldehyde + CoA-SH}$
3. $\mathrm{Mevaldehyde + NADPH + H+ \rightarrow Mevalonate + NADP+}$

Now we can proceed to the wonderful regulation of HMG-CoA reductase. The regulation takes place at transcription level, translation level, degradation and phosphorylation level. At gene level, the modulation is done by a transcription factor known as Sterol recognition element-binding protein

(SREBP). When the cholesterol level drops, the SREBP attaches to Sterol recognition element which is present on 5' side of the gene for enzyme reductase. This binding increases the transcription of mRNA for the enzyme. On the other hand, during high concentration of cholesterol in the cell, the SREBP is quickly destroyed in the nucleus, therefore transcription decreases.

Translation of mRNA for HMG-CoA reductase is inhibited by nonsterol derivatives of Mevalonate. The degradation of protein enzyme will obviously reduce the activity. High sterol levels bring structural changes in the membrane domain of the enzyme which accelerate the proteasome-mediated degradation. Cholesterol synthesis is an expensive process in terms of ATP. Therefore, the pathway should slow down when ATP is less. This is achieved by phosphorylation of enzyme by AMP-activated protein kinase. Phosphorylated form is inactive.

Regulation by Statins: Statins are the most commonly used lipid-lowering drugs also known as HMG-CoA reductase inhibitors. These drugs are prescribed in patients at high risk for cardiovascular diseases and hypercholesterolaemia. Statins are competitive inhibitors of HMG-CoA reductase. Recall that competitive inhibitors are the structural analogues of substrate of an enzyme which bind at active site. As the reaction is rate-limiting step of pathway, the effect is substantial (Fig. 7.4).

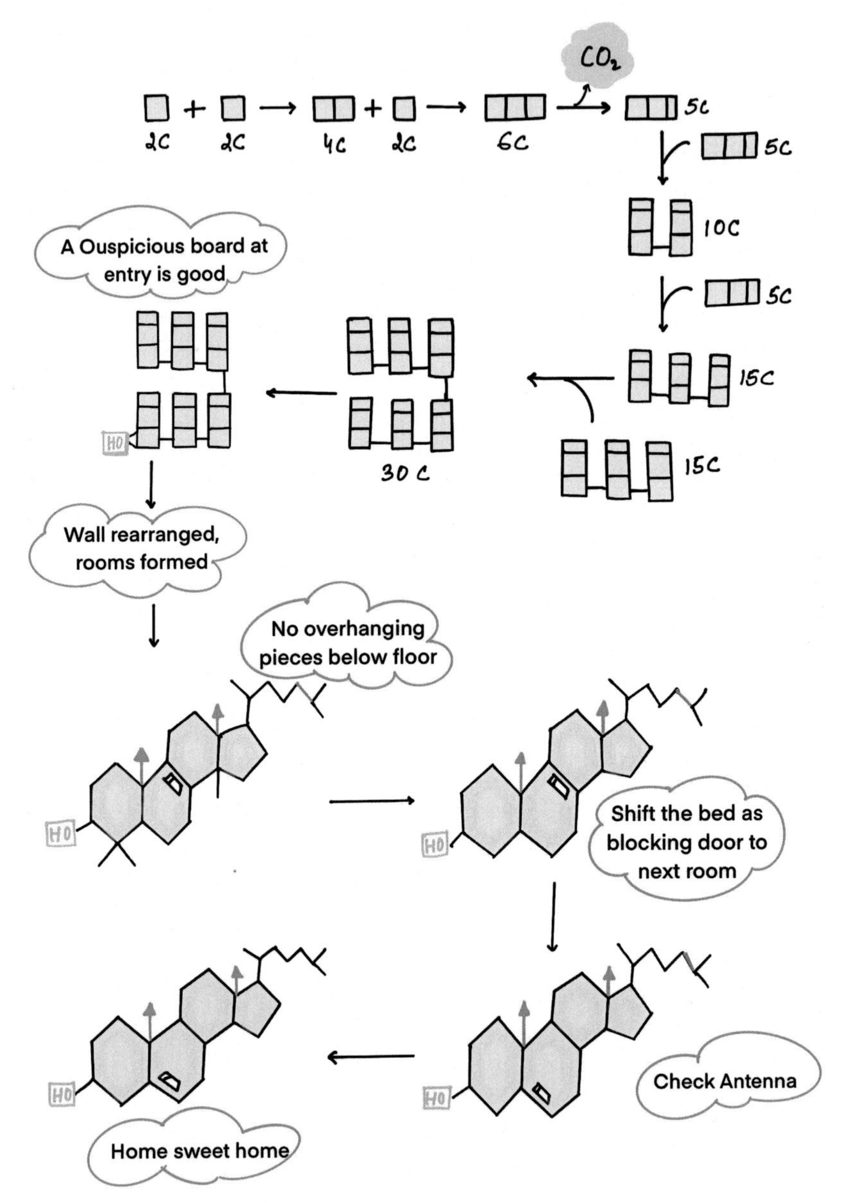

FIGURE 7.4 The queen bee narrates the story of making her honeycomb house.

Story of queen bee's honeycomb house

A queen bee decides to make her honeycomb house with three bedrooms and one kitchen. She started joining bricks of acetyl-CoA (2C) in the following manner.

2C + 2C = 4C
4C + 2C = 6C
6C − 1C = 5C
5C + 5C = 10C
10C + 5C = 15C
15C + 15C = 30C

With 30C, her material for house is complete. She starts with hanging an Ouspicious (auspicious) board at the entrance of house. Then she rearranges the walls in such a way that three hexagonal rooms and one pentagon kitchen are formed. She discards the extra hangings below the floor for the smooth look. Next, she moved her bed which was obstructing the way to the third room. Finally, when the house was ready, she corrected her TV antenna's settings (Table 7.1).

TABLE 7.1 Correlation between the queen bee's story and cholesterol synthesis.

Cholesterol synthesis	
WHAT BEE DOES	WHAT CORRELATES IN CHOLESTEROL PATHWAY
2C	Acetyl-CoA
4C	Acetoacetyl-CoA
6C	HMG-CoA
5C	Isoprenoid unit
10C	Geranyl
15C	Farnesyl
30C	Squalene
Ouspicious (auspicious) board at the entrance of house	Squalene is oxidised to Squalene 2,3-epoxide
She rearranges the walls	Cyclisation
She discards the extra hangings below floor	The methyl groups on C_{14} and C_4 are removed to form 14-desmethyl lanosterol
Next she moved her bed	Double bond at C_8–C_9 is subsequently shifted to
She corrected her TV antenna	Reduction of double bond of side chain

Chapter 8

Haem synthesis

Hemoproteins play a vital role in cellular functions from gaseous exchange to redox reductions. Biologically significant hemoproteins include haemoglobin, myoglobin, catalase, cytochrome c and cytochrome p450. Haem synthesis starts in the mitochondria with the condensation of Succinyl-CoA with Glycine amino acid, activated by Pyridoxal phosphate. Aminolevulinic acid (ALA) synthase catalyses this irreversible reaction forming an intermediate amino-ketoadipic acid. ALA Synthase (ALAS) is the rate-limiting enzyme of haem synthesis.

Two forms of ALA synthase are found: erythroid (ALAS2) and hepatic (ALAS1). ALA molecules enter the cytoplasm where their union in the presence of ALA dehydratase yields Porphobilinogen (PBG) and water molecule. ALAD is inhibited by lead, and haem synthesis is inhibited leading to anaemia. Four PBG molecules are joined by Uroporphyrinogen I synthase (PBG deaminase) as linear tetrapyrrole called Hydroxymethylbilane. Linear tetrapyrrole cyclises to form a ring known as Uroporphyrinogen III (UPG) with the participation of Uroporphyrinogen III synthase. Uroporphyrinogen III has one asymmetric side chain.

All acetyl groups of UPG are coverted to methyl groups by decarboxylation, and the coproporphyrinogen III (CPG) is generated. CPG is acted upon in the mitochondria by CPG oxidase which decarboxylates and oxidises two propionic side chains to vinyl groups. Protoporphyrinogen thus formed is further oxidised to protoporphyrins. Molecular oxygen is required for conversion of CPG to protoporphyrins. Iron is incorporated finally to generate haem.

Haem synthesis pathway is carried out by the bone marrow (major ontribution) and liver. Haem, the product of the pathway, regulates its synthesis by decreasing the synthesis of ALAS1 by negative aporepressor feedback.

Porphyrias are group of genetic disorders characterised by deficiency of enzymes of haem synthesis. Haem synthesis is affected in ALAS and ALAD deficiency (Fig. 8.1).

Sweet Biochemistry. DOI: https://doi.org/10.1016/B978-0-443-15348-8.00028-4

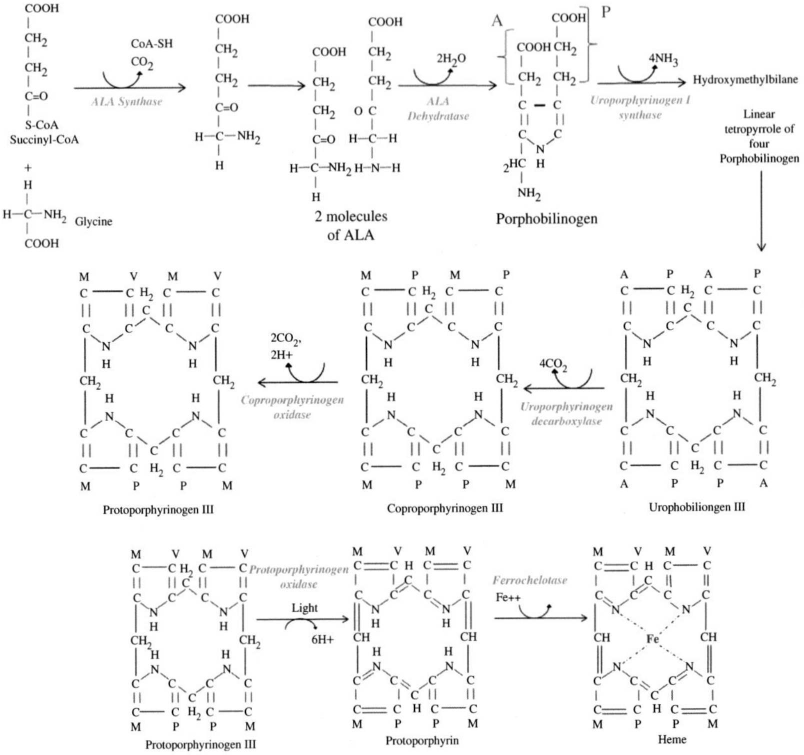

FIGURE 8.1 Haem synthesis steps.

Rate-limiting step of haem synthesis: aminolevulinic acid formation

ALAS is a member of α-oxoamine synthase subfamily of pyridoxal phosphate (PLP)-dependent enzymes with EC 2.3.1.37. PLP participates as cofactor with it for one-step condensation of glycine and succinyl-CoA to form delta-aminolevulinic acid. δ-Aminolevulinic acid (ALA) is a non-proteinogenic amino acid named so because amino group is attached on the C4 in molecule. ALAS structure is described as a tightly interlocked homodimer and monomer units have three domains (most of enzymes have three, just a coincidence??). N-terminal domain, C-terminal domain and catalytic domain. ALAS is located on the matrix side of the inner mitochondrial membrane. This enzyme is secreted as inactive precursor which is activated during translocation to its destination. PLP is affectionately (covalently) bound to a lysine residue at active site by a Schiff base.

In the reaction first substrate glycine amino acid enters and replaces the lysine residue at active site forming an external Aldimine. Then a H+ is lost stereospecifically from the alpha-carbon of PLP-bound glycine resulting in carbanion molecule. This is followed by a condensation of carbanion with Succinyl-CoA, the second substrate to produce an intermediate α-amino-β-ketoadipate aldimine still attached to the Enzyme. During it, the CoA-SH exits from the reaction. The intermediate with complex name is decarboxylated to yield ALA. But wait. This ALA is liberated by hydrolysis from the active site. PLP again joins lysine residue. PLP in ALA synthesis basically promotes H+ removal as electrophilic pyridinium ring works as an electron sink. The slowest part of process is ALA dissociation and hence governs the rate of reaction.

Regulation of aminolevulinic acid synthase

Turnover rate of ALAS in rat liver was found to be near 70 minutes. This crucial enzyme is regulated at multiple levels by multiple factors. Let's see the important ones. Haem, the final product of the pathway, decreases the activity of ALAS by repressing the ALAS1 gene transcription, disturbing the translation of ALAS1 mRNA and further destabilising it. Haem also inhibits the import of enzyme precursor from the cytoplasm to mitochondria. On oxidation, haem forms Haematin – an allosteric inhibitor of enzyme. In haematin, a hydroxyl group is linked to iron. If this hydroxyl is replaced by chloride anion, Hemin is formed which also works like Hematin. Intravenous Hemin can repress hepatic ALAS1. ALAS activity is also decreased by Glucose. This can be used as a simple treatment for diseases of haem pathway.

Aminolevulinic acid synthase activity in liver can be induced by steroids and some drugs by promoting ALAS1 gene expression. For equimolar concentration production of haem and globin in erythroid cells, identical transcriptional elements are present in genes of ALAS2 and globin. Haem production needs to be regulated with cellular iron levels, hence at translation stage, iron-responsive element of ALAS2 mRNA senses the iron levels, and translation is inhibited in case of insufficient iron levels. Isn't this a smart natural trick to coordinate whole process of haem synthesis? (Figs. 8.2 and 8.3).

Imagine you are a dance instructor preparing a group dance of HE-ME dance. Males are suCoA and females are glycine. You complete the dance in 8 instructions to dancers.

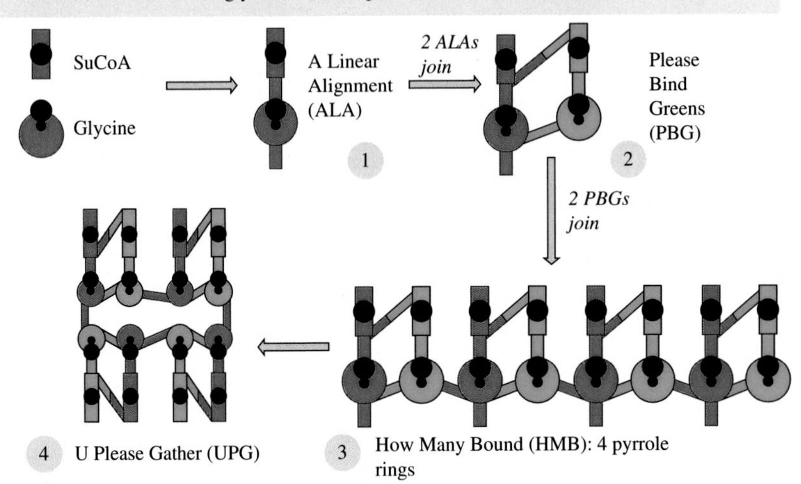

FIGURE 8.2 He-Me dance part 1.

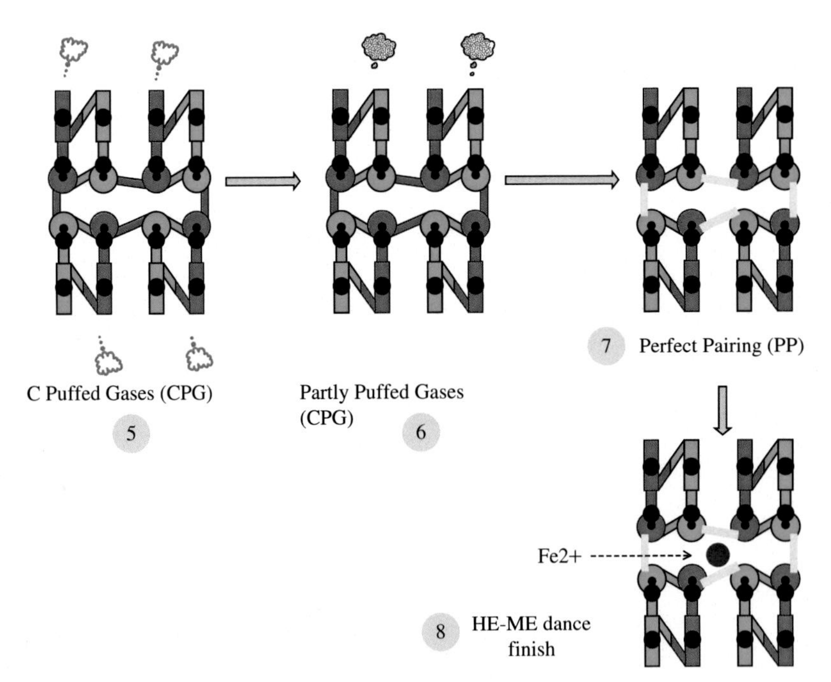

FIGURE 8.3 He-Me dance part 2.

Have a look at the Table 8.1 to correlate He-Me dance with Haem synthesis. Following is another simple trick to learn haem synthesis (Fig. 8.4).

TABLE 8.1 Correlation between the He-Me dance and haem synthesis.

Haem synthesis	
INSTRUCTIONS	WHAT CORRELATES IN HAEM SYNTHESIS PATHWAY
SuCoA	**Succinyl-CoA**
1. A linear alignment	ALA formation by succinyl-CoAand Glycine
2. Please bind greens	Porphobilinogen formation
3. How many bound	Hydroxymethylbilane formation
4. U please gather	Uroporphyrinogen
5. C (see) puffed gases	Coproporphyrinogen formation by decarboxylation
6. Partly puffed gases	Protoporphyrinogen formation by decarboxylation and oxidation
7. Perfect pairing	Protoporphyrin
8. He-Me	Haem formation

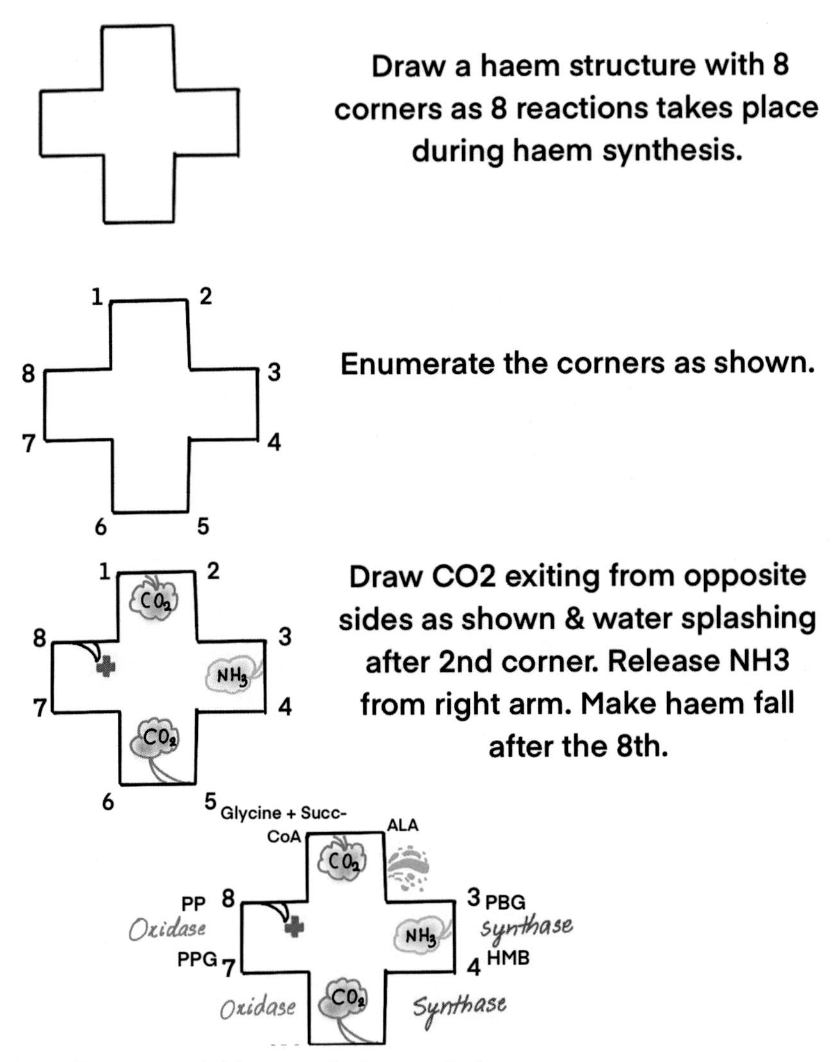

FIGURE 8.4 Second trick to remember haem synthesis.

Chapter 9

Porphyrias

Traditional recap

Porphyrias are inborn errors of metabolism caused by defects of enzymes involved in haem synthesis (Fig. 9.1). A characteristic finding of porphyrias is excretion of porphyrins and porphyrinogens in urine. Skin and nervous system are predominantly affected due to accumulation of porphyrins and porphyrin precursors while the liver, bone marrow and other organs may also be affected. Neuropsychiatric features and photosensitivity are noticed among the features. As discussed in the previous chapter, haem is an important prosthetic group for haemoproteins including cytochromes and haemoglobin participating in etc. and oxygen transfer. The synthesis of haem can occur in all cells, but the principal sites are the liver and bone marrow, explaining the involvement of these organs in porphyrias.

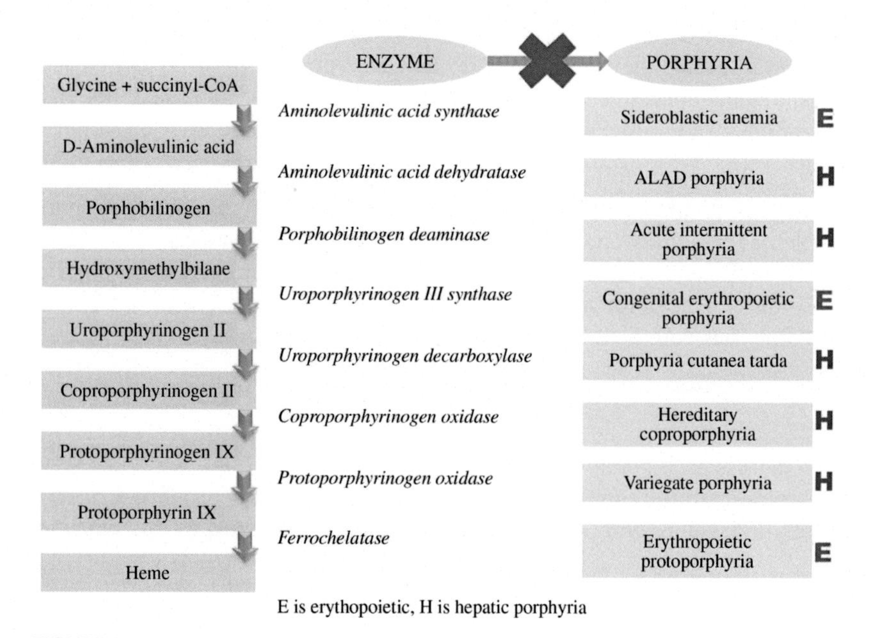

FIGURE 9.1 Porphyria and defective enzymes.

Sweet Biochemistry. DOI: https://doi.org/10.1016/B978-0-443-15348-8.00008-9

Manifestations of porphyria

Clinically the patient may present with acute porphyria, also referred to as acute hepatic porphyria, involving the nervous system. In this type, major presenting features are attack of abdominal pain, neuropsychiatric changes, constipation, hypertension, tachycardia and weakness of extremities (which can progress to whole body, requiring respiratory support). Neuropsychiatric features range from insomnia, anxiety, hallucinations, delusions to seizures. These episodes may take days to weeks to recover. Delta-aminolevulinic acid (ALA) dehydratase deficiency porphyria, acute intermittent porphyria, hereditary coproporphyria and variegate porphyria are four acute porphyrias. Don't be scared of these complex terms. When defects will become clear, these terms will look simpler.

Cutaneous porphyria is divided into blistering and nonblistering type. Blistering porphyrias are correlated with chronic skin changes (sunlight exposure causes blisters, scars and hyperpigmentation) as compared with nonblistering ones. In the latter, the sunlight exposure leads to more acute, painful skin lesions, affecting the behaviour of the patient also, but the skin changes are relatively less chronic.

All porphyrias are caused due to genetic mutations coding for the enzyme of haem synthesis except one. Effect of mutation may be silent, only aggravated in hepatic porphyria by external factors such as some medications, alcohol intake, dietary changes, hormone administration or chemicals. Most common porphyria – Porphyria Cutanea Tarda (special one) is not due to mutation but external factors. Autosomal dominant inheritance is observed in porphyrias except or aminolevulinic acid dehydratase (ALAD) deficiency porphyria and congenital erythropoietic porphyria.

Porphyrias are also categorised on the basis of the tissue affected in three groups:

1. Hepatic porphyria including acute intermittent porphyria, variegate porphyria, hereditary coproporphyria and porphyria cutanea tarda.
2. Erythropoietic porphyria including hereditary erythropoietic porphyria, erythropoietic porphyria.
3. Porphyrias with both hepatic and erythropoietic features.

Microview

Let's discuss each porphyria type one by one.

Aminolevulinic acid synthase (ALAS) **deficiency** (erythroid form) results in aanemia with low haemoglobin and erythrocytes, as this enzyme catalyses the first step of haem synthesis. X-linked dominant erythropoietic porphyria is caused due to two ALAS2 gene mutations. This mutation results in over-active ALAS elevating the ALA production in RBC. ALA is converted to porphyrins which cross the bloodstream to reach the skin and other tissues. Thus photosensitivity appears in such patients. Another disease X-linked sideroblastic anaemia is a result of ALAS gene mutation on chromosome X (Fig. 9.2).

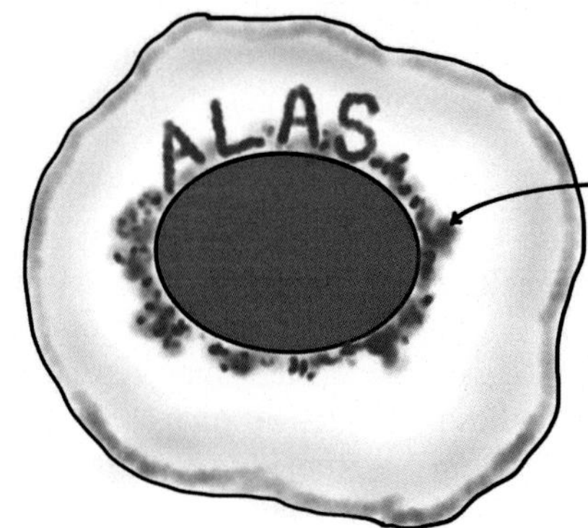

FIGURE 9.2 X-linked sideroblastic anaemia and Acute Intermittent porphyria mnemonic.

ALAD deficiency porphyria presents with abdominal pain and neuropsychiatric symptoms along with increased ALA and coproporphyrin III. Abdominal pain is accompanied by vomiting and constipation. Peripheral neuropathy, motor nerves involvement and psychosis can occur during acute attacks. Severity of symptoms correlates inversely with the residual enzyme activity. It is a hepatic form of porphyria caused due to mutation in ALAD gene transmitted in autosomal recessive manner. Stress, alcohol, fasting, infection, some drugs, dehydration and use of oestrogen are risk factors for acute attack. Lead is another metal which can inhibit this enzyme.

Uroporphyrinogen (UPG) I synthase deficiency/acute intermittent porphyria Acute intermittent porphyria (AIP) (hepatic form) is characterised by increased porphobilinogen (PBG) and d-ALA urinary excretion along with abdominal pain and neuropsychiatric features. Symptoms appear mostly in adulthood, and the gender more commonly affected is female. Mutation in the gene for the enzyme porphobilinogen deaminase (second name hydroxymethylbilane synthase) leads to AIP. Inheritance is autosomal dominant. Photosensitivity is not a feature of AIP because significant porphyrin synthesis does not take place.

UPG III synthase congenital erythropoietic porphyria The defect of UPG III synthase is known as congenital erythropoietic porphyria (CEP). Uroporphyrin I is produced in excessive amounts which deposits in the tissue and is also excreted via urine. This is a very rare autosomal recessive disease, beginning in childhood. In mild cases, the blisters may resemble porphyria cutanea tarda (PCT). Hypertrichosis (abnormal increased hair growth) is also seen in some cases. Uroporphyrin I is formed in large quantities, thus it accumulates in the tissues (in the skin, uroporphyrin being photoactive damages the skin) and gets excreted in urine. However, the most important point is that due to deficiency of type III isomer, the negative feedback for type I porphyrin formation is further reduced. Extremely high porphyrin levels in RBC cause haemolysis. A pathognomonic pink to dark red colour is observed in the teeth, bones and urine. Photosensitivity is present. Uroporphyrins demonstrate a strong red fluorescence under UV light (Figs. 9.3 and 9.4).

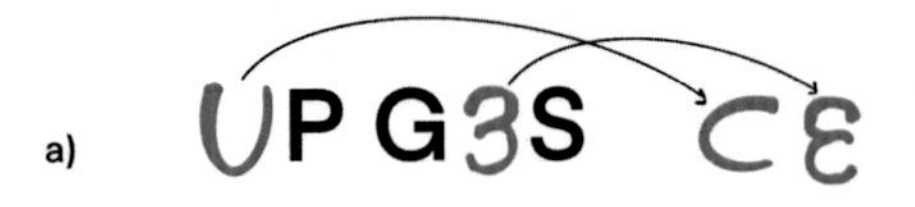

FIGURE 9.3 Congenital erythropoietic porphyria mnemonic.

(A) Porphyria cutanea tarda

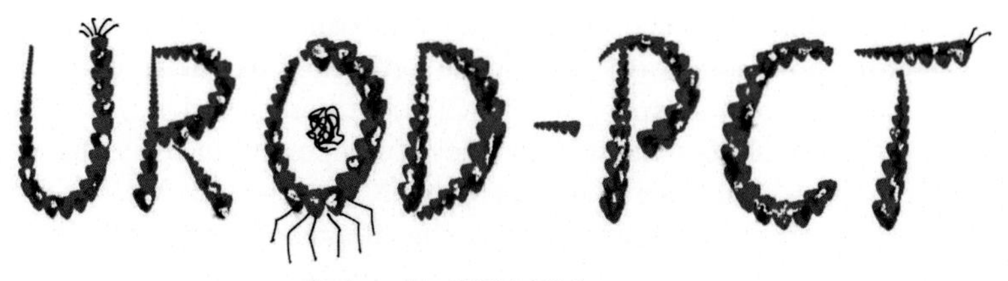

(Induced by HBV, HIV)

(B) Hereditary coproporphyria

(C) Variegate Porphyria

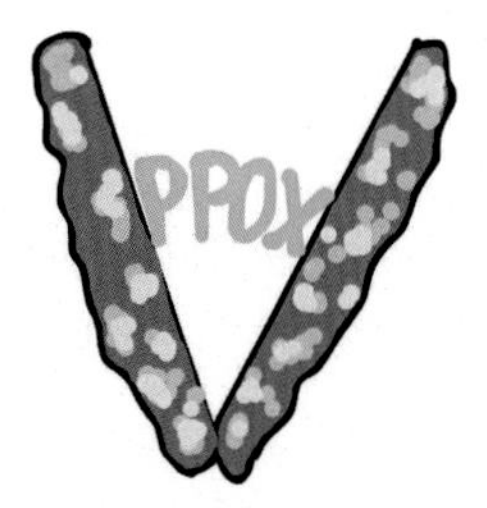

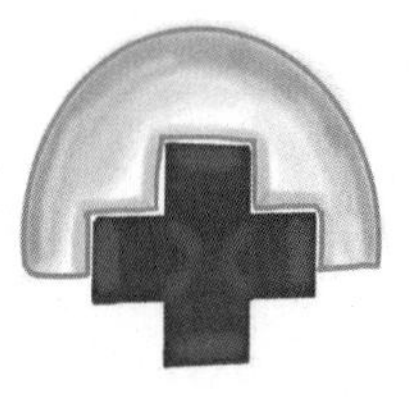

FIGURE 9.4 Mnemonics for porphyria cutanea tarda, hereditary coproporphyria, variegate porphyria and ferrochelatase deficiency.

UPD decarboxylase deficiency porphyria is also called **PCT**. Patients with this disease have photosensitivity, liver disease, elevated hepatic iron stores, and increased uroporphyrin. Alcohol intake and oestrogen therapy are well-known precipitating factors. This is the most common and most treatable porphyria with an autosomal dominant transmission (Fig. 9.4).

CPG oxidase deficiency or hereditary coproporphyria includes photosensitivity, abdominal pain and neuropsychiatric features. Urinary ALA, PBG and Coproporphyrin III and faecal coproporphyrin III increased (Fig. 9.4).

Protoporphyrinogen (PPG) oxidase deficiency or variegate porphyria (VP) resembles coproporphyrinogen (CPG) oxidase deficiency, but the difference is that in CPG oxidase deficiency Coproporphyrin III is raised, while in PPG oxidase deficiency protoporphyrin IX is increased. Mutation responsible for VP may not manifest during the lifetime of a person. This is the second most common acute porphyria inherited as autosomal dominant. Episodes of VP resemble attacks of AIP. Blisters of porphyria cutanea tarda may be confused with VP (Fig. 9.4).

In **ferrochelatase deficiency or erythropoietic protoporphyria**, protoporphyrin is elevated in the reticulocytes, RBC, plasma and faeces. Enzyme is deficient in the reticulocytes in the bone marrow. Cells with high protoporphyrin exhibit red fluorescence when fluorescent light is exposed. The inheritance is autosomal dominant, but the expression varies. This is a mild disease with photosensitivity being a major complaint. Elevated protoporphyrin is taken up by the liver for secretion in the bile, meanwhile getting damaged. In this disorder, urinary porphyrin levels are normal (Fig. 9.4).

Why photosensitivity and neuropsychiatric symptoms are observed in porphyrias?

Porphyrins have a unique cyclic tetrapyrolle structure and a deep purple colour. The conjugated ring with alternating single and double bonds results in complex absorption spectrum. At around 400 nm, porphyrins provide a very strong absorption band. The band becomes weaker at 500–650 nm.

Photosensitisation reaction can be divided into two categories. Type I – a direct transfer of electron takes place from photosensitiser to substrate. Type II – energy from light is passed to molecular oxygen creating singlet O_2 and reactive oxygen species.

Protoporphyrin accumulated in dermal capillaries is excited by light to triplet state. This promotes singlet oxygen formation which causes cell damage. Porphyrins interact with iron on exposure of photons and induce generation of reactive oxygen species (ROS). In porphyria cutanea tarda, uroporphyrin accumulated in the skin along with photoactivation of complement system causes release of proteases from mast cells in the dermis layer. This leads to dermal–epidermal separation presenting as skin fragility and vesicles. ROS also stimulates the pain receptors transient receptor potential ankyrin1 (TRPA1) and transient receptor potential cation channel subfamily V member 1 (TRPV1) promoting secretion of neuropeptides causing excruciating burning pain often described as 'ice-cold fire on the skin'.

Pain is reported by patients with porphyrias even within minutes of visible light or neon light exposure

The dark-coloured (dark brownish) urine, the clinical feature that derives the term 'porphyrus', may be unremarkable because ALA and PBG are colourless and are most observed after exposure of the voided urine to light, leading to the oxidative reaction of porphobilinogen to uroporphyrin and porphobilin (uroporphyrin-like pigments correlate with the typical urine colour).

Administration of antioxidants is found helpful as it ameliorates the oxidative stress

Neuropsychiatric features

To understand the cause of neurological involvement, one should remember the enzymes for which haem is acting as cofactor, namely haemoglobin, hepatic cytochrome P450, myoglobin, mitochondrial respiratory chain cytochromes, catalases, peroxidases, microsomal cytochrome b 5. It is vital to know the basic pathways where these are involved. By just this, the havoc the derangements in the pathways will cause may be deduced. But here are some mechanisms compiled for reading:

Aminolevulinate (ALA) is neurotoxic due to structural resemblance of ALA with neurotransmitter GABA. GABA participates in regulation of peristalsis and muscle tone regulation in the intestine.

Low availability of haem will directly reduce mitochondrial oxidative phosphorylation. Elevated ALA can also compromise this. Problems in oxidative phosphorylation are responsible for dysfunction of Na + -K + ATPase channel in axonal membrane.

At high ALA levels, autooxidation of ALA which is cytotoxic damages the mitochondrial membrane and increases ROS production leading to neurovisceral features and cancer in the liver.

Porphyrins disturb protein oxidation and aggregation processes

High ALAS1 activity may deplete pyridoxal phosphate, adding further to secondary sensory axonal neuropathy

Amino acid metabolism especially of tryptophan, glycine, acetylcholine and noradrenaline is hampered. There is increased production of neuroactive metabolites from tryptophan in porphyria (Fig. 9.5). Below given Fig. 9.5 summarises the mnemonics of various porphyrias.

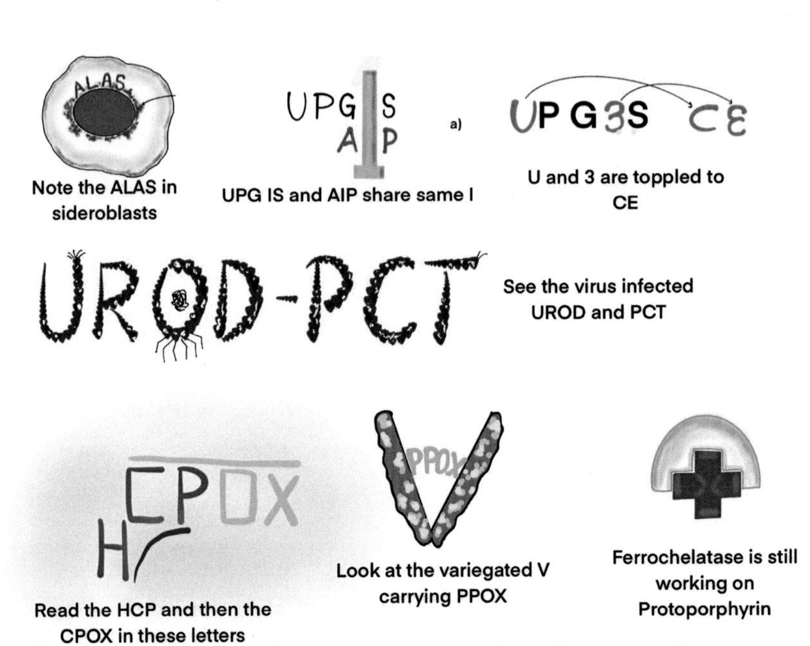

FIGURE 9.5 Summary of porphyria mnemonics.

Chapter 10

Urea cycle

Urea cycle or ornithine cycle helps to excrete two harmful gases ammonia and carbon dioxide out of the body. Ammonia is highly toxic to the central nervous system and needs to be eliminated from the body. Steps of this cycle take place in the mitochondria and cytoplasm. Liver is the main organ synthesising urea along with the kidney to a lesser extent. Alpha nitrogen of amino acids is excreted in the form of urea.

Amino acids transfer their amino groups to alpha ketoacids such as alpha-ketoglutarate (α-KG) by transamination reaction catalysed by transaminases with pyridoxalphosphate as coenzyme. Alpha-KG converts to glutamate during this process. Glutamate dehydrogenase acts upon glutamate and releases ammonia along with α-KG.

Chemically urea is H_2N-CO-NH_2. One amino group comes from NH_4^+ and second from aspartate while carbon is derived from bicarbonate. First step of urea cycle is condensation of CO_2 and NH_3 to carbamoyl phosphate by carbamoyl synthetase 1 (CPS 1), using 2ATP in the mitochondria. Carbamoyl phosphate is a high-energy compound. CPS1 enzyme is activated by N-acetylglutamate (NAG), and this is different from CPS II which participates in pyrimidine synthesis. NAG is synthesised by NAG synthetase which is activated by arginine and glutamate.

Ornithine enters a mitochondrion through a transporter. Enzyme ornithine transcarbamoylase (OTC) transfers carbamoyl group to ornithine yielding citrulline which comes out of the mitochondria for further steps. L-aspartate is condensed with citrulline to form argininosuccinate by argininosuccinate synthetase to contribute second nitrogen of urea molecule. This step requires adenosine triphosphate (ATP) and releases adenosine monophosphate (AMP). Arginino- succinate is lysed by argininosuccinase to release arginine and fumarate. Urea is produced by hydrolysis of guanidino group of arginine along with regeneration of ornithine. Four ATP are consumed for one urea molecule synthesis.

Urea cycle is linked to tricarboxylic acid (TCA) cycle as aspartate donates one amino group of urea and fumarate released in urea cycle can enter TCA cycle (Fig. 10.1).

Sweet Biochemistry. DOI: https://doi.org/10.1016/B978-0-443-15348-8.00015-6

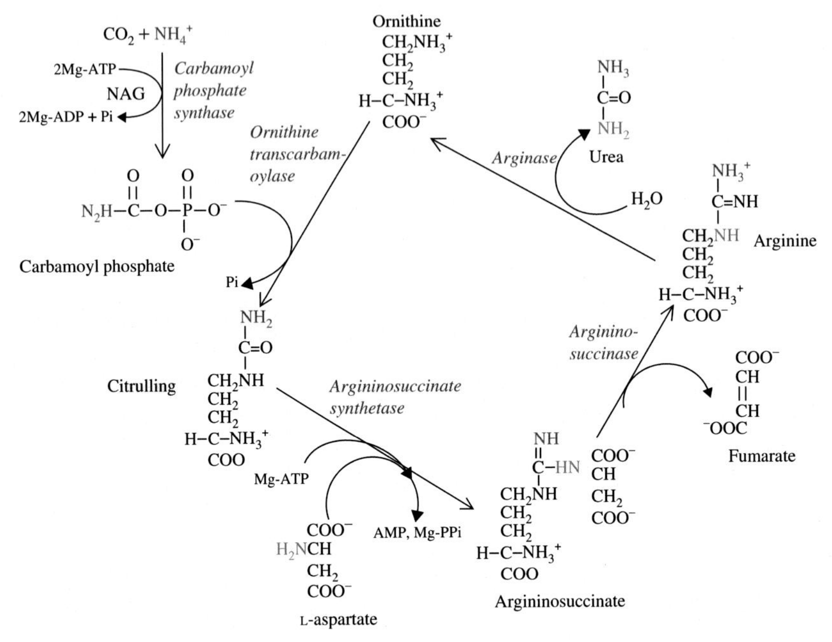

FIGURE 10.1 Steps of urea cycle.

Carbamoyl phosphate synthetase 1: an unique enzyme

Carbamoyl phosphate synthetase 1 (CPS 1), a mitochondrial ligase with E.C. 6.3.4.16, condenses bicarbonate and ammonia to form carbamoyl phosphate, using ATP. Structurally, CPS1 is a 1500-residue multidomain enzyme complex. Striking feature of this enzyme is the absolute dependence on NAG for activity. Dependence is to the extent that the enzyme is non-functional in the absence of NAG. Deficiency of NAG cannot be clinically distinguished from deficiency of enzyme itself. The synthesis of NAG becomes crucial because CPS1 catalyses the rate-limiting step by siphoning ammonia and carbondioxide towards urea formation. Ammonia, being a strong base in nature, exists as ammonium ion NH_4^+ in aqueous medium predominantly, but CPS1 chooses only ammonia as its substrate and not the ionic form.

Are there other modulators also for carbamoyl phosphate synthetase 1?

When NAG enjoys obligatory association with CPS1, Carbamoyl Glutamate doesn't want to be left behind and hence is another activator. Other relative molecules of NAG: Glutamate and alpha-ketoglutarate are competitive binders of NAG synthase. Product of pathway Mg-ADP brakes the enzyme activity.

Enzyme catalysing the synthesis of NAG from acetyl-CoA and Glutamate is NAG synthase. Positive allosteric modulator for NAGS is Arginine, and the product NAG exerts negative allosteric modulation. CoA-SH depletion reduces NAG synthesis and consecutively urea synthesis because CoA-SH is needed for Acetyl-CoA formation. NAGS and CPS-1 are competitively inhibited by CoA derivatives.

So it is clear that reaction of CPS-1 is regulated by multiple factors, not just NAG.

It is a wonder, how does NAG affects CPS-1? CPS-1 possesses two binding sites for ATP. One ATP is spent on activation of bicarbonate resulting in carboxy-phosphate, attached to the enzyme. Carboxy-phosphate unites with ammonia to form Carbamate, and inorganic phosphate is removed. One more ATP is required for forming carbamoyl-phosphate, and energy released is used in separating it from the enzyme. Consumption of two ATP in this three-step set turns this reaction irreversible. The presence of anhydride bond in carbamoyl group of carbamoyl-phosphate allows the molecule to have high transfer potential.

This explains the reaction mechanism. Now comes the role of NAG. When NAG binds CPS-1 at C-terminal domain, a dramatic long-range conformational change is triggered in both phosphorylation domains. On binding of ATP, a remodelling is induced, which stabilises the active conformation. In this form, a 35 angstrom long tunnel appears for migration of carbamate from first phosphorylation site to second site where carbamoyl-phosphate is produced.

On closer observation, NAG levels correlate with nitrogenous burden in the body. NAG comes from glutamate and has short half-life in vivo. Moreover, the affinity of NAGS for glutamate is very low (indicated by high Km). But from where glutamate comes? It is derived from trapping ammonia. The body cannot deplete all glutamate for ammonia detoxification because low glutamate levels may start unnecessary protein breakdown. Therefore, NAG should be formed only when excess ammonia (or glutamate) is produced, not otherwise (Table 10.1).

NAG also increases affinity of CPS-1 for its activators K^+ and Mg^{++}. These mechanisms highlight the role of NAG in CPS-1 catalysis (Figs. 10.2 and 10.3) (Table 10.1).

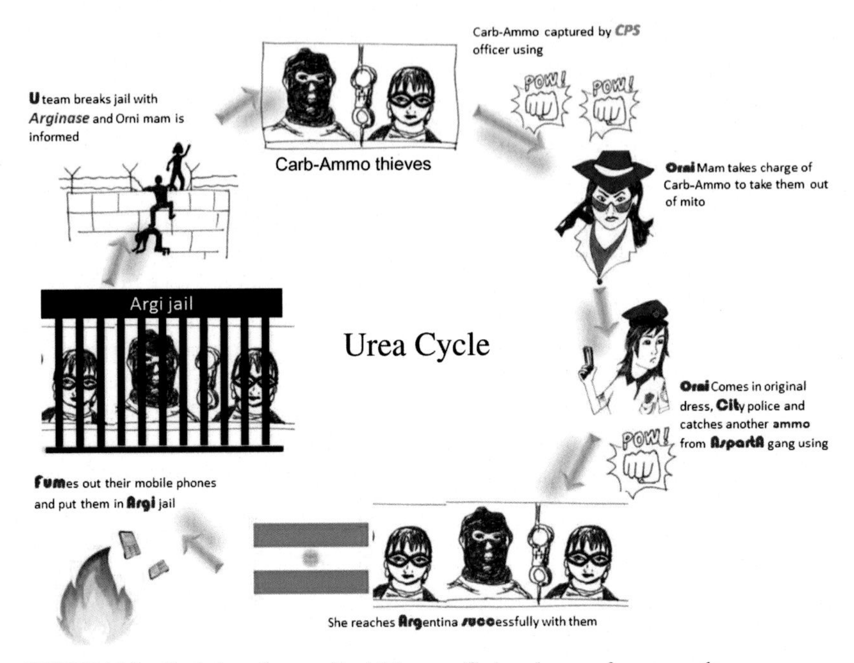

FIGURE 10.2 Depiction of story of bad thieves – Carb and ammo for urea cycle.

Urea cycle story

Two bad thieves: carb and ammo were causing lot of nuisance in holy city of human by forming a secret team carbamoyl. An honest CPS officer captured the two using lots of energy. Orni Mam takes the two in custody to cautiously take them out of mito.

Orni Mam comes in her original dress of City Police. Here she catches another sister thief ammo from asparta gang after a fight. She reaches Argentina Successfully with them. Here she takes out their phones and set them on fumes so that they cannot contact anyone. The three thieves were sent to Argi jail. Inside jail they formed a U team and escapes the jail bars with Arginase gel. Orni Mam is informed again about the same. Thus the story of run and catch continues.

FIGURE 10.3 Story to remember urea cycle.

TABLE 10.1 Urea cycle story

What happens in story	What happens in urea cycle
Carb-Ammo	Carbon dioxide-ammonia
CPS officer	Carbamoyl phosphate synthetase
Punch image with POW	ATP
Orni Mam	Ornithine
City police	Citrulline
AspartA gang	Aspartate
Argentina successfully	Argininosuccinate
Fumes out	Fumarate
Argi Jail	Arginine
U Team	Urea

Chapter 11

Urea cycle disorders

Urea cycle disorders (UCDs) are related to defects of enzymes involved in urea cycle. It affects all age groups.

All enzymes of this pathway give rise to UCD. Carbamoyl phosphate synthetase I (CPS I) found in mitochondria is used for urea cycle. Liver is the main site of urea cycle (kidney being a minor contributor). CPS I catalyses the rate-limiting step. N-acetyl glutamate (NAG) is an allosteric activator of CPS I. NAG is synthesised from acetyl-CoA and glutamate. In high concentrations of ammonia, alpha-ketoglutarate is converted to glutamate and hence is not available for the Krebs cycle. Therefore, ATP production reduces by impairing the Krebs cycle. This accounts for high toxic features of ammonia like irreversible brain damage, coma and death.

Hyperammonemia, encephalopathy and respiratory alkalosis are the principal signs and symptoms of UCDs. A characteristic feature of UCD is that ammonia toxicity is highest when the block is at the initial 1 or 2 stages (Fig. 11.1). Products after citrulline are comparatively lesser toxic, and moreover some ammonia has been consumed upto this reaction; hence symptoms are most severe in hyperammonemia I. Ammonia and glutamine are elevated in deficiency of CPS, OTC, ASS and ASL.

Sweet Biochemistry. DOI: https://doi.org/10.1016/B978-0-443-15348-8.00019-3

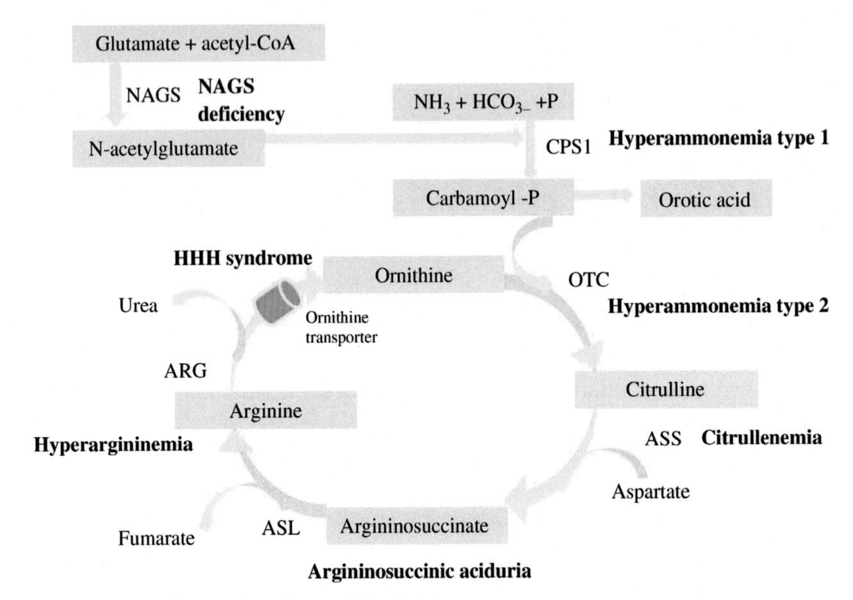

FIGURE 11.1 Urea cycle and urea cycle disorders.

Patients of UCD complain of vomiting, lethargy, irritability, severe mental retardation and aversion to high-protein foods. Neonates may suffer from additional apnoea and hypothermia at high ammonia concentrations. Members of UCDs have similar presenting features, but at level of enzyme block the substrates of enzyme increase and the concentration of products falls. Transporters like ornithine transporters and citrin deficiency are also included in UCDs.

Infections, high-protein diets, starvation, physical exertion, drugs and surgery can elevate ammonia levels and aggravate symptoms of patients. Fig. 11.2 shows the mechanisms behind CPS1 deficiency.

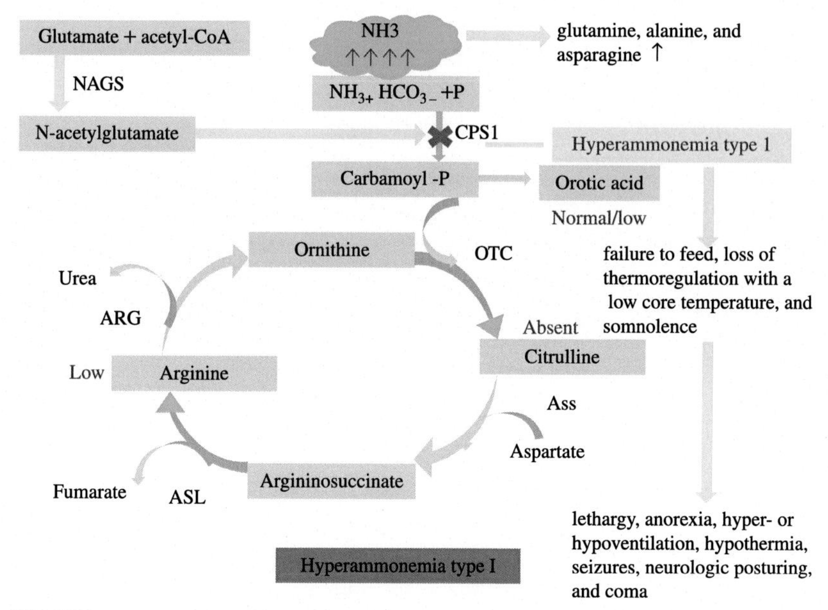

FIGURE 11.2 CPS deficiency. The yellow arrows indicate the level reactions can go.

Presentation of *N*-acetyl glutamate synthase defect resembles CPS1 defect. Fig. 11.3 Ornithine transporter disturbances are also included in urea cycle disorders. Fig. 11.4 Ornithine Transcarbamylase deficiency also known as Hyperammonemia type 2 is a X-linked disease. Fig. 11.5 Citrullinemia type 1 is Argininosuccinate synthetase deficiency, in which Citrulline rises upto 10-fold times in body fluids. Fig. 11.6 Next enzyme deficiency Argininosuccinate lyase leads to excretion of Argininosuccinate in urine. Fig. 11.7 Arginase deficiency causes elevation of Arginine in body fluids like blood and cerebrospinal fluid. Fig. 11.8 A summary of important features of different UCDs is given in the last diagram. Fig. 11.9 UCDs revolves around decreasing ammonia production by restricting protein intake. This minimises rising of ammonia levels which ameliorates brain damage. Levulose is administered to promote acidic gut environment which converts NH_3 to NH_4^+. This stops absorption of ammonia from gut. Compounds like benzoate and phenylacetate are helpful by binding amino acids; hence, it lowers ammonia production.

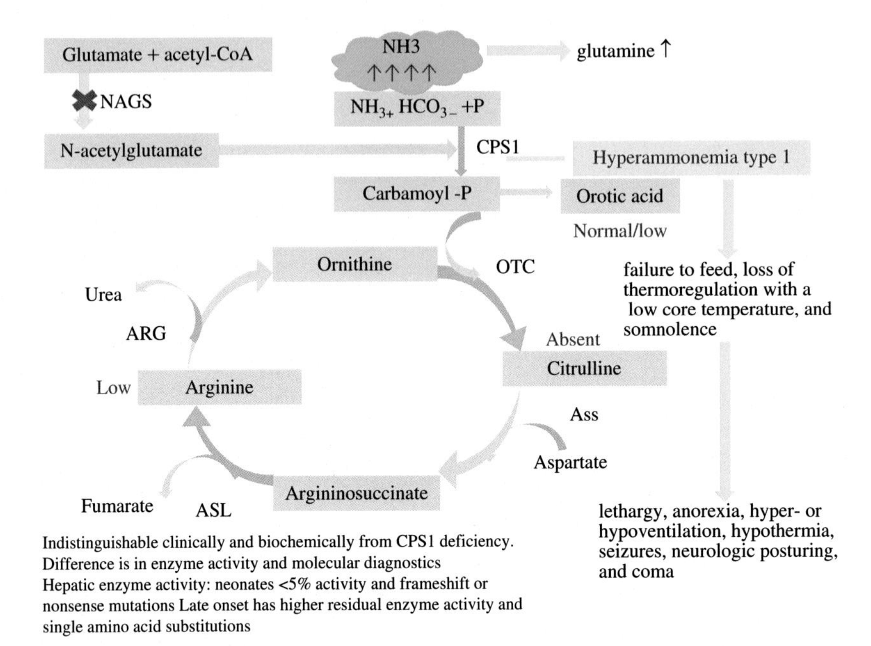

FIGURE 11.3 NAGS deficiency.

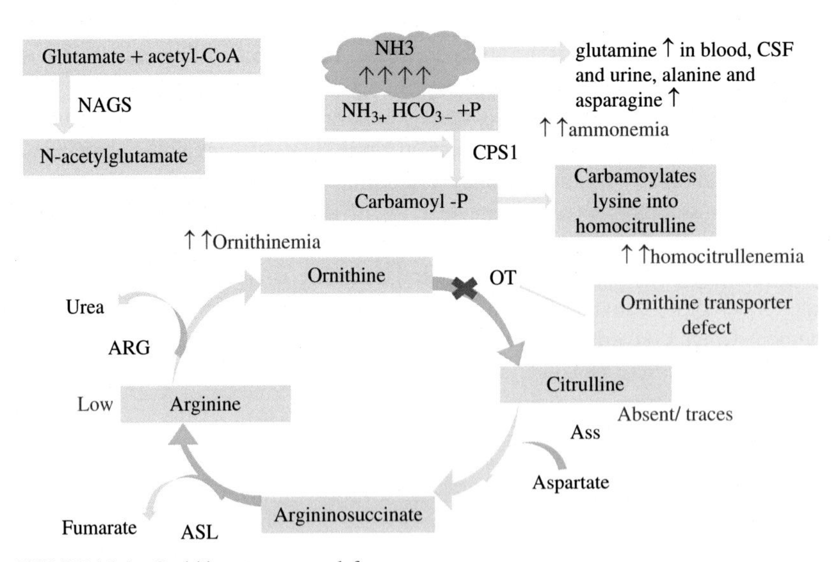

FIGURE 11.4 Ornithine transporter defect.

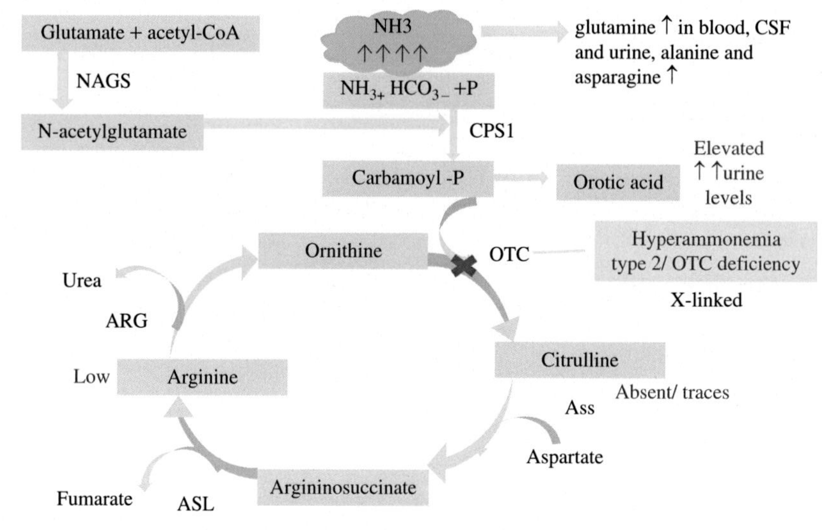

FIGURE 11.5 Ornithine transcarbamoylase deficiency.

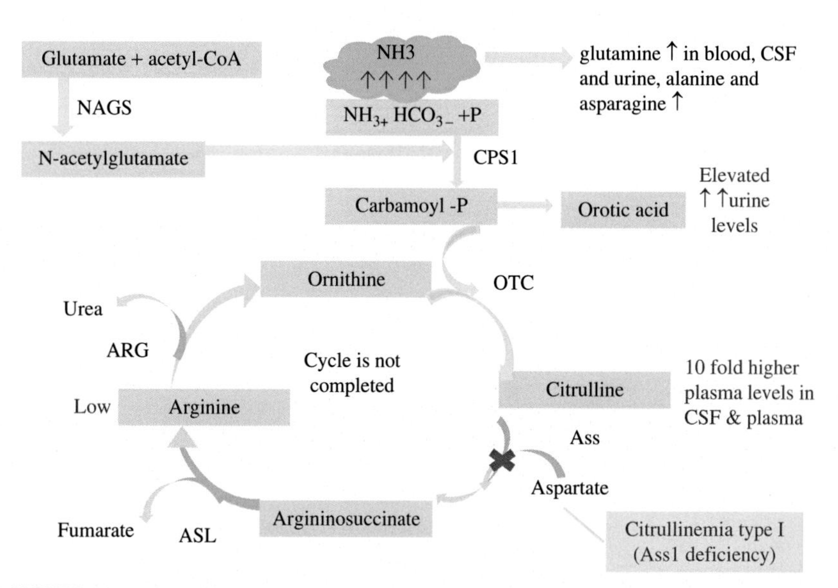

FIGURE 11.6 Citrullinemia type 1

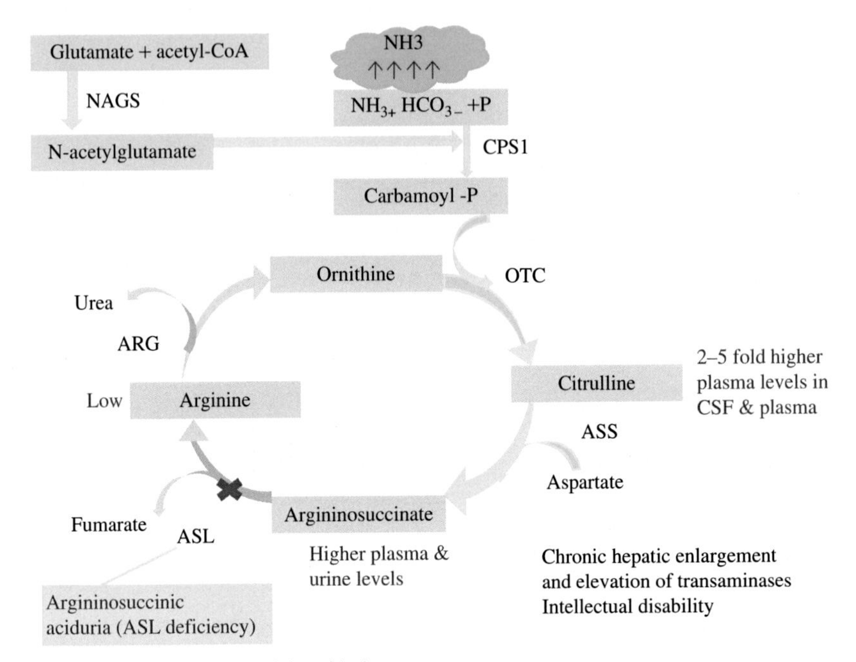

FIGURE 11.7 Argininosuccinic aciduria.

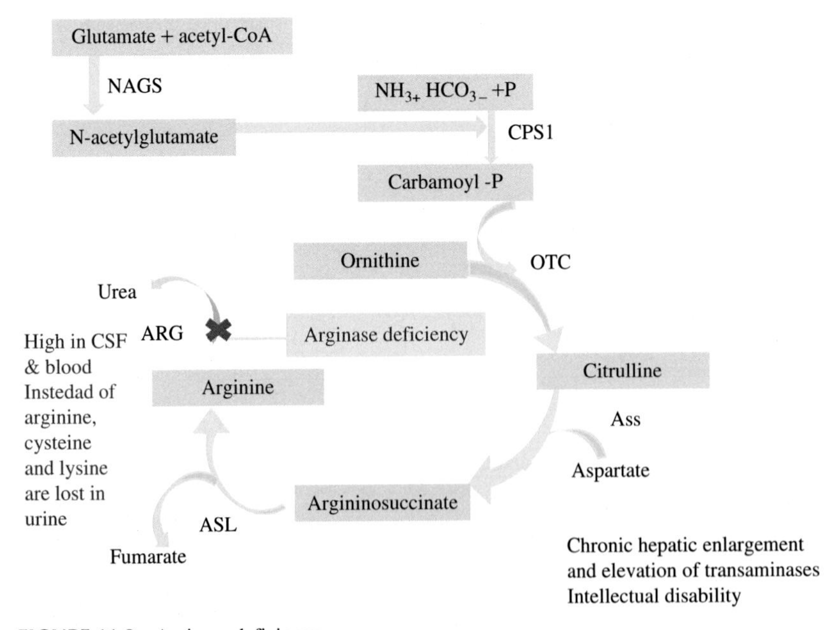

FIGURE 11.8 Arginase deficiency.

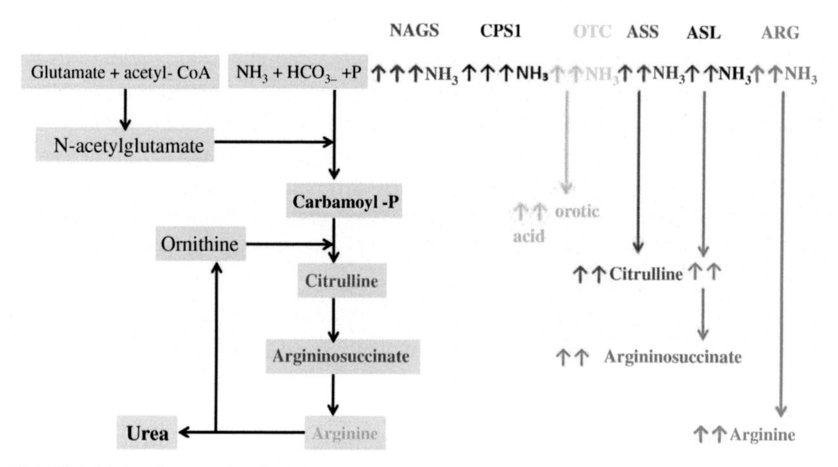

FIGURE 11.9 Urea cycle disorders summary. On the top enzymes are written on right side. Then with the same colour lines the level up to which pathway works is shown. Main accumulating substances are written in each disorder.

Why ammonia is so toxic to brain?

Let us understand the mechanisms responsible for the deleterious effects of ammonia.

Ammonia is an irritant gas, NH3. So it is uncharged. The presence of nitrogen allows it to act as a weak base with pK of 9.2–9.8. At 7.4 physiological pH, almost 99% of NH3 is present in the form of NH_4^+. Usually, the blood ammonia levels vary between 15 and 45 μg/dL, and at more than 1 mM concentration, ammonia becomes toxic to mammalian cells. When the level of NH3 rises, mostly which is due to either overproduction or less elimination, it starts to spread alkalinity everywhere. Due to the small size and neutral charge, it can easily travel across the lipid membranes of cell and organelles. So it elevates the pH of cytoplasm as well as organelles. Now, this is a big problem because many enzymes require an optimum acidic pH to work, like those present in lysosomes. Therefore, the activity of proteases in lysosomes and function of Golgi vesicles are impaired.

Hyperammonemia promotes GABA (an inhibitory neurotransmitter) release by activating peripheral benzodiazepine receptors situated on astrocytes.

Ammonia causes toxic levels of glutamine accumulation in astrocytes, the main victim. Cell swelling and cell death ensue. Astrocytes are the site where glutamate and ammonia combine to form glutamine. When this safety mechanism is overused, glutamine level becomes high and glutamate (an excitatory neurotransmitter) low, and another problem arises. The activity of malate-aspartate shuttle is lowered leading to fall in pyruvate:lactate ratio in astrocytes. Elevated ammonia level also inhibits decarboxylation of alpha-ketoglutarate (intermediate in TCA cycle) in astrocytes and neurons. This further inhibits pyruvate dehydrogenase complex, resulting in bioenergetic failure. In rat brain studies, ammonia was found to inhibit all enzyme complexes etc., adding to the damage. It also disrupts H+ gradient required for ATP synthesis. Cerebral ATP depletion along with rise in intracellular calcium lead to increase in potassium levels outside the cell. This is another factor for cell death.

Overactivation of NMDA receptors by ammonia in acute stage initiates oxidative or nitrosative stress in astrocytes and neurons. These NMDA receptors after chronic ammonia exposure get downregulated which impairs cGMP synthesis.

Recent research has proposed that ammonia directly inhibits EAAT-1 and EAAT-2 transporters participating in the disposal of glutamate from neuronal synapse. Derangements in NO synthesis and signal transduction pathways are also observed in hyperammonemia.

Ammonia increases the transport of aromatic amino acids which are precursors of neurotransmitters like serotonin and dopamine. High serotonin causes sedation, while high dopamine presents as motor impairment.

Developing brain of children is more prone to ammonia toxicity. The extensive CNS damage in the form of cortical atrophy, demyelination and ventricular enlargement is responsible for seizures, cerebral palsy, cognitive impairment, drowsiness and coma. If not managed timely, death occurs. For example, newborns with enzyme defects in urea cycle may appear normal at birth but become significantly ill within 2 days.

Chapter 12

Glycogen storage disorders

Glycogen is a highly branched storage form of glucose in animals. Glucose units are joined by alpha 1,4 glucosidic linkages and branching carries alpha 1,6 glucosidic bond. Liver and muscle are two major areas of glycogen storage (muscle being the larger source). Liver by glycogenolysis releases free glucose in blood, while muscle lacks glucose-6-phosphatase, so the glucose-6-P in muscle enters glycolysis and forms pyruvate which enters gluconeogenesis.

Glycogen synthesis: Glycogenin protein core glucosylates itself at its tyrosine residue with UDP-glucose in the beginning. Glycogen synthase then forms glucosidic bonds between C-1 of the glucose of UDPGlc and C-4 of a terminal glucose residue of glycogen. Branching enzyme creates a branch by alpha 1,6 bond, when the chain is 11 glucose units long. This process of lengthening and branching continues to yield a highly branched glycogen molecule.

Glycogen breakdown: Glycogen degradation starts with addition of phosphates to glucose from non-reducing ends by glycogen phosphorylase, the rate-limiting enzyme. Glucose-1-P is released sequentially until four glucose residues remain on the branch. Transferase enzyme shifts trisaccharide to the other branch leaving behind a glucose joined with alpha 1,6 bond. Debranching enzymes hydrolyses the alpha1,6 glycosidic bond to release free glucose. Glycogen phosphorylase continues its action on the straight chain.

Glycogen synthase and glycogen phosphorylase are regulated by phosphorylation. CAMP increased by different factors activates cAMP-dependent protein kinase. This enzyme phosphorylates phosphorylase kinase and activates it. Phosphorylase kinase in turn activates glycogen phosphorylase enzyme.

Glycogen storage diseases

Glycogen storage diseases (GSDs) are also known as Glycogenosis or Dextrinosis. They are a group of inherited disorders resulting from defective glycogen synthesis or degradation leading to accumulation of glycogen in liver, muscles and other tissues. Glycogen which is deposited may be normal

Sweet Biochemistry. DOI: https://doi.org/10.1016/B978-0-443-15348-8.00010-7

in structure or abnormal. Accumulation causes hepatomegaly (kidney enlargement in some), hypoglycemia and muscle weakness. In GSDs, Glucose-6-phosphate from muscles enters PPP pathway to elevate purine synthesis leading to hyperuricemia (Fig. 12.1).

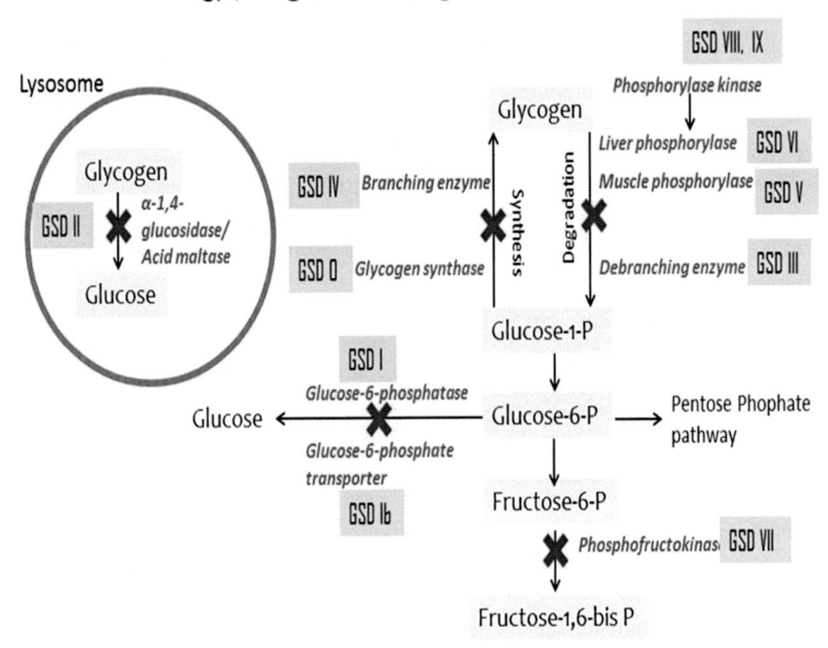

FIGURE 12.1 Glycogen storage disorders along with enzyme blocks.

Mechanisms for features of glycogen storage disorders

We should first understand the features of Type I GSD – Von Gierke's disease, then it is very easy to understand what happens in all types (Type I to Type X) that are summarised in Table 12.1.

In Von Gierke's disease, the enzyme glucose-6-phosphatase is deficient (most common cause). This enzyme is present in liver, kidney and intestine. Due to inactivity of this enzyme, the phosphate cannot be separated from glucose. Glucose-6-phosphate is the principal source of free glucose production as glycogenolysis as well as gluconeogenesis are finally producing this molecule. Both of these pathways maintain the blood glucose levels in between the meals or during exercise. Inability to release required free glucose results in severe hypoglycaemia and short-term fasting intolerance, frequently observed in interprandial state. There are many consequences of low

TABLE 12.1 Summary of glycogen storage disorders with enzymes and clinical features.

Glycogen storage disorders summary		
Type	**Enzyme defect**	**Clinical features**
Type 0	Liver glycogen synthase	Hypoglycemia, hyperketonemia, mild hepatomegaly and early death
Type Ia (Von Gierke's disease)	G-6-phosphatase deficiency	Hypoglycemia, massive hepatomegaly, renomegaly, hyperuricemia, ketosis, lactic acidosis, hyperlipidemia and short stature
Type Ib	Endoplasmic reticulum G-6-P Transporter defect	Type Ia features with doll-like faces with fat cheeks, short stature, and protuberant abdomen, bleeding tendency, recurrent infections
Type II (Pompe's disease)	Lysosomal acid maltase	Lysosomes laden with glycogen, diminished muscle tone, cardiorespiratory failure and enlarged tongue
Type III (Cori's disease)	Liver and muscle debranching enzyme	Hypoglycemia, cirrhosis, hepatomegaly, muscle and cardiac involvement (milder course than Type 1), and characteristic branched polysaccharide deposits
Type IV (Anderson's disease)	Branching enzyme	Hepatosplenomegaly, liver cirrhosis, failure to thrive, death usually before 5 years and glycogen with few branch points deposited
Type V (McArdle syndrome)	Ms. phosphorylase	Exercise intolerance, myalgia (muscle pain), muscle contractures, hyperCKemia and myoglobinuria
Type VI (Her's disease)	Liver phosphorylase	Hepatomegaly, hypoglycemia, ketosis and moderate growth retardation
Type VII (Tarui's disease)	Muscle, RBC phosphofructokinase 1	Muscle pain and cramps, and decreased endurance
Type VIII	Liver phosphorylase kinase	Mild hepatomegaly and mild hypoglycemia
Type IX	Liver and muscle phosphorylase kinase	Hepatomegaly, mild hypoglycemia, short stature and muscle weakness
Type X	cAMP-dependent protein kinase A	Muscle aches or cramping following strenuous physical activity, recurrent myoglobinuria and kidney failure

blood glucose levels. Infants may suffer from convulsions as brain is heavily dependent on glucose for energy requirements. Administration of epinephrine or glucagon fails to raise the blood glucose concentration. As expected, low glucose signals insulin concentration fall in blood, while glucagon rises. This hormone combination promotes lipolysis and glycogenolysis. Glycogen breakdown further exacerbates the problem.

Phosphorylated glucose is trapped inside the cell as there are no transporters for it in the plasma membrane. Its presence stimulates the glycolysis, ending up in piling of pyruvate and lactate in extrahepatic tissues. Lactate is sent to liver to enter gluconeogenesis pathway.

Therefore, the body is trying to handle hypoglycaemia by glycogen breakdown and generating new glucose, but it is not sufficient. Only endogenous source for free glucose in little amount is action of debranching enzyme.

Glucose-6-phosphate is also shunted to other pathways where it can be consumed like pentose phosphate pathway and PRPP synthesis. This amplifies purine synthesis, that on breakdown generates uric acid. High uric acid level predisposes the patient to Gout and other hyperuricemia complications. Hyperlactacidemia also hampers with uric acid excretion.

As mentioned before, low glucose levels urge the dense energy source biofuel – lipids, to sustain energy requirements of various organs. Let me explain the mechanism. Glycolysis ends in pyruvate which forms acetyl-CoA by pyruvate dehydrogenase complex reaction. Accumulation of acetyl-CoA indicates high energy state and promotes lipogenesis. Acetyl-CoA is the precursor of fatty acids and cholesterol. On the one hand, it amplifies the synthesis of these lipids, while on the other hand acetyl-CoA forms malonyl-CoA which prohibits rate-limiting step of oxidation of fatty acids by inhibiting carnitine palmitoyl transferase I, thereby decreasing the breakdown of fatty acids. Hypoglycaemia favours flux of fatty acids from adipose tissue to the liver for energy production.

So how such patients can be managed. Main triggering factor hypoglycaemia is prevented by frequent daytime feedings and continuous nocturnal intra-gastric feeding with high glucose formula. This allows better growth of child and decreases liver size also. Uncooked cornstarch is one of the good choices for feeding.

Story mnemonic of glycogen storage disorders

Glycogen storage disorders are very difficult to remember. I have represented this in the form of commonest warrior story. First enjoy the story in Fig. 12.2 and then try to learn it by heart. Correlate the words of story with GSD as per Table 12.2.

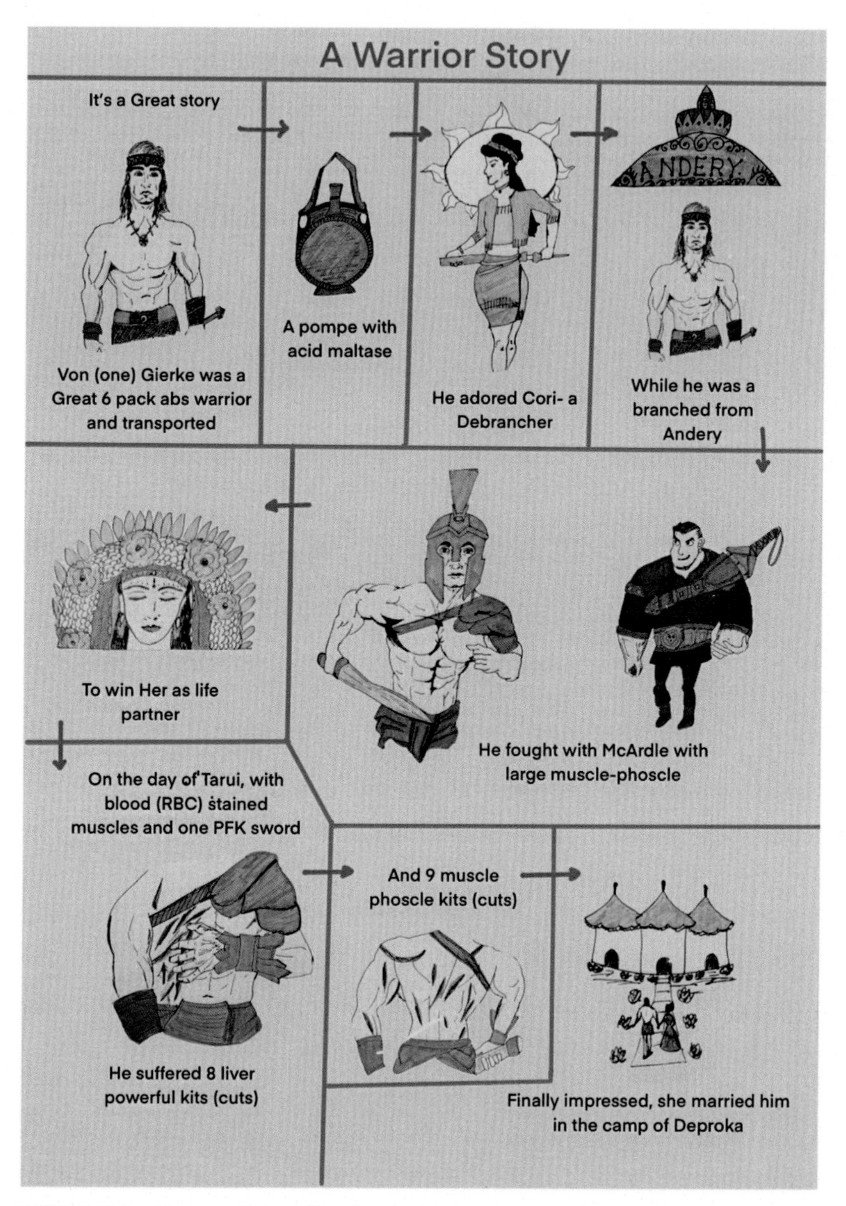

FIGURE 12.2 Glycogen Storage disorders depicted as the story of warrior Von Gierke.

TABLE 12.2 Correlation of story with GSD (glycogen storage disease).

What relates to GSDs in story?			
Type	**Story part correlating**	**Disease**	**Enzyme defect**
0	It's a Great Story	Glycogen synthase defect	Glycogen synthase
Ia	Von (one) Gierk was a Great 6 Pack abs warrior	Von Gierke's disease	G-6-Ptase deficiency
Ib	And transported		ER G-6-P transporter defect
II	A pompe with acid maltase	Pompe's disease	Lysosomal acid maltase
III	He adored Cori – a debrancher	Cori's disease	Liver and muscle debranching enzyme
IV	While he was a brancher from Andery	Anderson's disease	Branching enzyme
V	He fought with McArdle with large muscle phoscle	McArdle syndrome	Muscle. phosphorylase
VI	To win Her as Life Partner	Her's disease	Liver phosphorylase
VII	On the day of Tarui, with blood (RBC) stained Muscles and one PFK sword	Tarui's disease	Muscle, RBC phosphofructokinase 1
VIII	He suffered eight Liver Powerful Kuts (cuts)		Liver phosphorylase kinase
IX	And nine Muscle Phoscle Kuts (cuts)		Liver and muscle Phosphorylase kinase
X	Finally impressed, she married him in camp of deProKA		cAMP-dependent protein kinase A

Chapter 13

Mucopolysaccharidoses

Glycosaminoglycans (GAGs) are also known as mucopolysaccharides due to their presence in mucosa. Chemically, these GAG molecules are long, highly negatively charged and unbranched heteropolysaccharides composed of repeating disaccharide monomer.

Components of this disaccharide unit are as follows:

1. Amino sugar: N-acetylgalactosamine (GalNAc) or N-acetylglucosamine (GlcNAc).
2. Uronic acid: glucuronate (GlcA) or iduronate (IdoA) (Fig. 13.1).

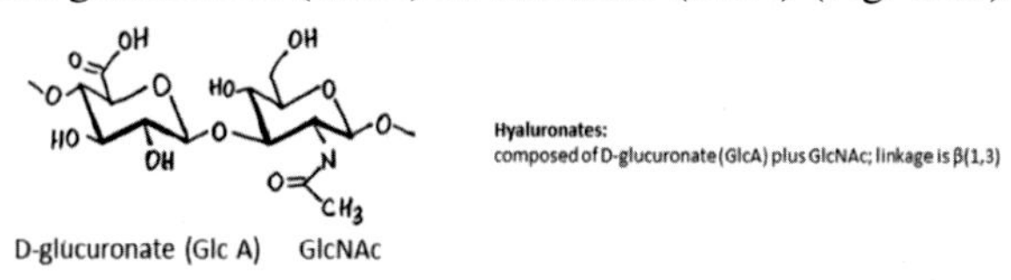

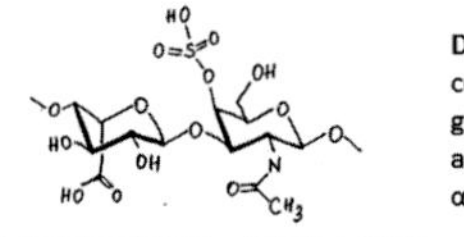

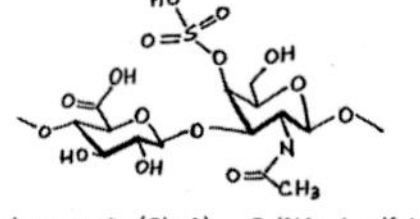

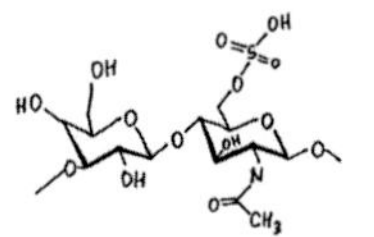

FIGURE 13.1 Mucopolysaccharides molecular structure.

Sweet Biochemistry. DOI: https://doi.org/10.1016/B978-0-443-15348-8.00006-5

Depending on hexosamine units linked alternatively with uronic acid, GAG can be classified into either glucosaminoglycans including heparin, hyaluronic acid and heparin sulphate or galactosaminoglycans like chondroitin sulphate or dermatan sulphate.

Sulphate and hydroxyl groups are present imparting strong negative charge and extended conformation. These properties make their solution very viscous. GAG may covalently bind with proteins to form proteoglycan. Carbohydrates may contribute >95% of proteoglycan's weight. Proteoglycans act as excellent lubricants and shock absorbers due to their water retention, great volume and strength. GAG are present in bone, cartilage, tendons, cornea, skin, joints and connective tissue, so these are affected by deposition.

The Syndecans are transmembrane cell surface proteoglycan of heparan sulphate, having a great role in cell–surface interactions. Syndecans can also connect with fibronectin, collagens, antithrombin III and thrombospondin, etc.

Earlier it was thought that mucopolysaccharidoses are a group of inherited lysosomal enzyme deficiency only which impairs the catabolism of GAGs. Emerging role of GAGs and lysosomes is however challenging this hypothesis. Derangements in myriad cellular processes like abnormal membrane structure affecting vesicle fusion and trafficking, secondary substrate storage, disturbance in autophagy, oxidative stress, mitochondrial dysfunction and dysregulated signalling pathways have been implicated in the pathophysiology of mucopolysaccharidoses (Table 13.1).

It is time to delve in the diseasesin little detail. Meanwhile for learning these disorders read this incidence. Hurler, Hunter Sir and Scheie were asked to show their identification cards (ID-Iduronidase).

All the three were not having their IDs. Hurler said innocently he had no idea where his ID was.

Hunter Sir could not hear the question properly due to deafness. Scheie said very smartly: IDs can be fake why worry so much about them?

Notice: Hurler and Scheie are wearing black goggles due to corneal opacity. Hunter is having mark of X on his shirt because of X-linked inheritance. He is called Sir to indicate sulphatase. Hurler and Hunter are both mentally retarded. Skeletal deformity is present in all three syndromes (Fig. 13.2).

TABLE 13.1 Mucopolysaccharidoses summary.

Mucopolysaccharidoses summary (MPS)			
Type	**Inheritance**	**Deficient enzyme**	**GAG (glycosaminoglycan) accumulated**
MPS I (Hurler, Hurler–Scheie, Scheie syndrome)	Autosomal recessive	α-L-iduronidase	Dermatan sulphate, Heparan sulphate
MPS II (Hunter syndrome)	X-Linked	Iduronate-2-sulphatase	Dermatan sulphate, Heparan sulphate
MPS III (Sanfilippo syndrome)	Autosomal recessive		Heparan sulphate
Type A		Heparan sulphatase	
Type B		N-acetylglucosaminidase	
Type C		Acetyl-CoAGlucosaminide acetyl Transferase	
Type D		Acetyl glucosamine-6-sulphatase	
MPS IV (Morquio syndrome)	Autosomal recessive		
Type A		Galactose-6-sulphatase	Dermatan sulphate
Type B		Beta-galactosidase	Chondroitin sulphate
MPS VI (Maroteaux–Lamy)	Autosomal recessive	Arylsulphatase B	Dermatan sulphate
			Chondroitin sulphate
MPS VII (Sly syndrome)	Autosomal recessive	Beta-glucuronidase	Dermatan/heparan
			Chondroitin sulphate
MPS IX	Autosomal recessive	Hyaluronidase	Hyaluron

FIGURE 13.2 Reactions of Hurler, Hunter and Scheie on asking for ID.

Hurler disease

Enzyme called lysosomal alpha-L-iduronidase which degrades mucopolysaccharides is deficient in Hurler syndrome:

- It is the most severe type of MPS I.
- Deposition of heparan sulphate and dermatan sulphate takes place.
- Affected individual is normal at birth but exhibits the following features:

Severe mental retardation, progressive deterioration, hepatosplenomegaly, corneal clouding and retinal degeneration, deafness and enlarged tongue, dwarfism with hunched back, nerve compression, restricted joint movements, and cardiac failure due to infiltration of heart.

Scheie syndrome: Mucopolysaccharidoses I

Enzyme called lysosomal alpha-L-iduronidase which degrades mucopolysaccharides is deficient.

It is the least severe form of Hurler syndrome.

Deposition of heparan sulphate and Dermatan sulphate takes place.

Common features include normal intelligence, less progressive deterioration and approximate normal lifespan, corneal clouding, restricted joint movements and valvular heart disease.

Hunter syndrome: Mucopolysaccharidoses II

Hunter syndrome is caused by deficiency or absence of the enzyme iduronate-2-sulfatase (I2S) in lysosones. It is the only MPS which is X-linked. Excess of heparan sulphate and dermatan sulphate are accummulated in organs like:

- Lungs: limited lung capacity.
- Heart valves: decreased cardiac function.
- Liver, spleen: hepatosplenomegaly.
- Joints: joint stiffness.
- Brain: mental retardation.

Other common features include abdominal hernias, repeated ear and respiratory tract infections, deafness and coarse facial features.

Sanfilippo syndrome: Mucopolysaccharidoses III

Here you can see the stages of a Sanfilippo child. First stage is normal baby, second stage is slow development shown by getting no 'A' grades in study, third stage is hyperactive, angry child and fourth stage is wheelchair-bound child. Different types of Sanfilippo syndrome along with enzyme involved are mentioned below figures correlated with mnemonics. Sanfilippo syndrome is due to accumulation of Heparan Sulphate (Fig. 13.3).

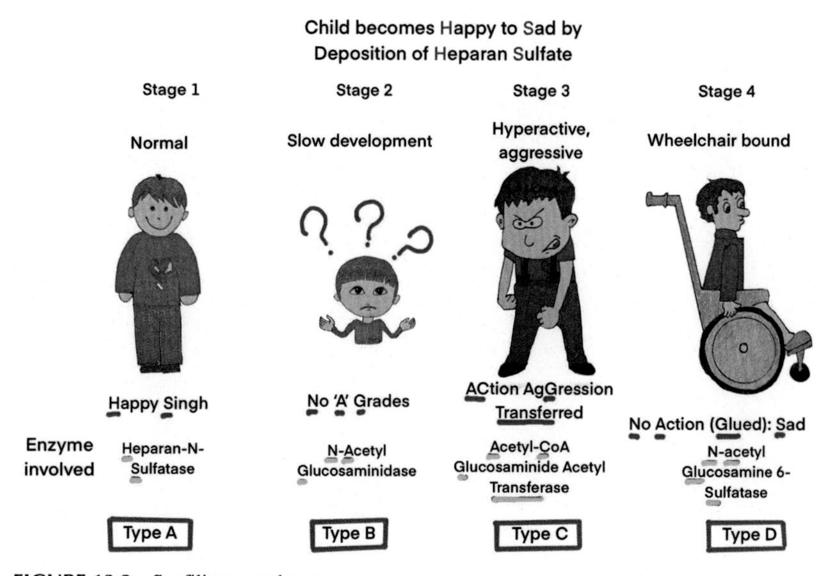

FIGURE 13.3 Sanfilippo syndrome.

Sanfilipo syndrome is caused by deficiency or absence of the enzyme involved in degradation of heparan sulphate in lysosomes. It is an autosomal recessive disease. Heparan sulphate is a GAG present in extracellular matrix and cell surface. Four enzymes defects have been discovered namely:

1. Heparan N-sulfatase.
2. N-acetylglucosaminidase.
3. Acetyl-CoA glucosaminide acetyl transferase.
4. N-acetyl glucosamine-6-sulphatase.

Patients of Sanfilippo syndrome present with mild facial dysmorphism, stiff joints, hirsuteness and coarse hair like other MPS patients. But in this syndrome, the progression follows a pattern of four stages. Initially, the child appears to be normal, but in few months to years development slows. Progressive motor disease, severe dementia and poor performance in studies are common complaints in second stage. In third stage, behavioural disturbances appear with hyperactivity, aggressiveness, pica and disturbances in sleep. In the last stage, child becomes unable to walk, wheelchair bound along with seizures and swallowing difficulties. These patients usually live upto early twenties or younger.

Morquio syndrome: Mucopolysaccharidoses IV

In Morquio syndrome, Keratan sulphate is piled up in tissues because of deficiency of either N-acetyl-galactosamine-6-sulphate sulfatase in Morquio syndrome type IVA or β-galactosidase in Morquio syndrome type IVB. Catabolism of chondroitin 6-sulphate is also affected. KS is abundantly present in cornea and cartilage; hence, these tissues are majorly affected. Type IVA is more severe form than IVB. Affected individuals present with mental retardation, skeletal deformity like scoliosis, kyphosis and rib flaring, epiphyseal dysplasia, corneal opacity and odontoid hypoplasia (Fig. 13.4).

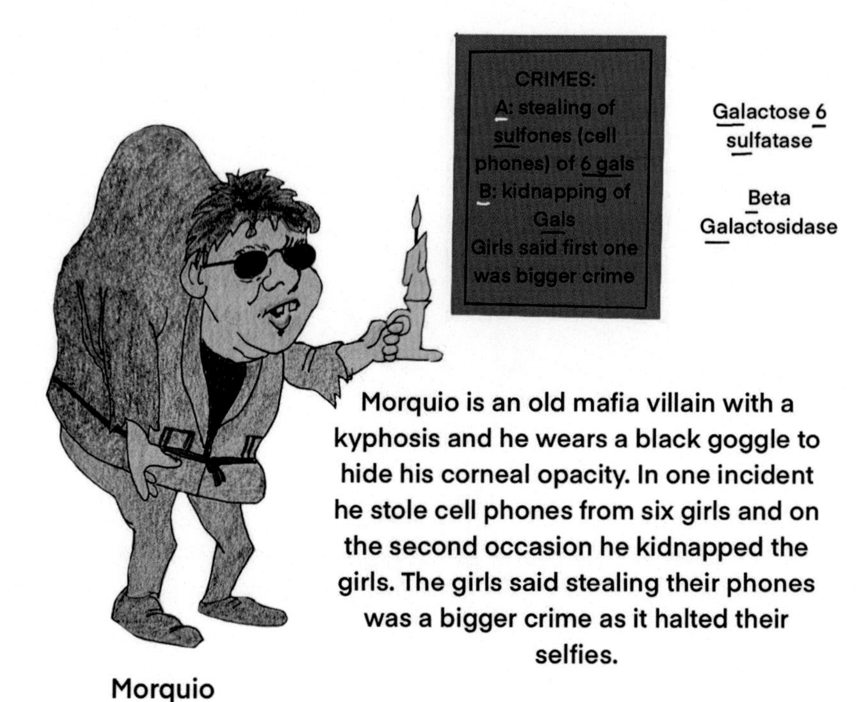

FIGURE 13.4 Morquio syndrome.

Maroteaux–Lamy syndrome: Mucopolysaccharidoses VI

Maroteaux–Lamy is the deficiency of N-acetylgalactosamine 4-sulphatase. GAG deposited is dermatan sulphate. Characteristic features include dysostosis multiplex, short stature, corneal clouding, deafness, dural thickening and pain caused by nerve compression developmental delay. Heart defects are also observed in patients (Fig. 13.5).

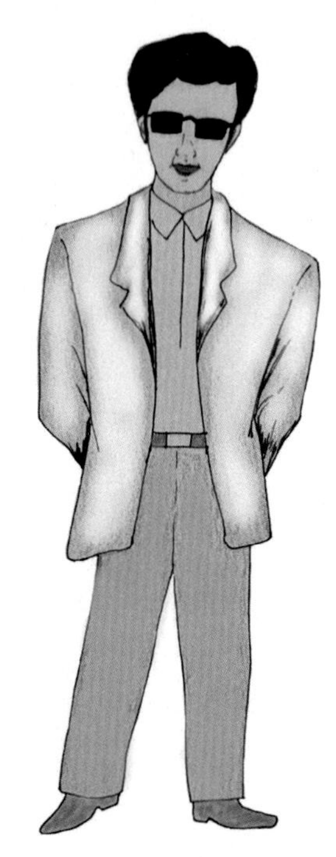

FIGURE 13.5 Maroteaux–Lamy syndrome.

Sly syndrome: Mucopolysaccharidoses VII

Sly syndrome is a lysosomal storage disease caused by deficient beta-glucuronidase leading to accumulation of dermatan sulphate, heparan sulphate and chondroitin sulphate. Major clinical manifestations include the following:

Dysmorphic facial features such as coarse face, macrocephaly and frontal prominence are present, corneal clouding or opacity, hepatosplenomegaly, skeletal features – dysostosis multiplex, kyphosis, hernias and vascular complications arise due to connective tissues involvement, mental retardation and developmental delay, and wide spectrum of presentations including hydrops fetalis in severe form and a neonatal form with jaundice is also present (Fig. 13.6).

FIGURE 13.6 Sly syndrome.

Chapter 14

Lipid storage disorders/ sphingolipidoses

Sphingolipidoses are member of lysosomal storage disorders in which enzymes of sphingolipid catabolism are absent. In these disorders, the sites of sphingolipid catabolism like lysosomes of phagocytes like histiocytes or macrophages in bone marrow, liver and spleen are affected. As sphingolipids like ceramide play a crucial role in membrane structure of neurons as well as regulate rate of neuronal growth and differentiation, any disturbance in these pathways lead to neurological impairments.

Sphingolipid catabolism is carried out by hydrolases present in lysosomes, and the structural components are removed sequentially. Structural components of sphingolipids have been described in previous chapter. Various enzymes participating in this pathway are galactosidases, glucosidases, neuraminidase, hexosaminidase, sphingomyelinase (a phosphodiesterase), sulphatase and ceramidase (a amidase). Irreversible reactions catalysed by these enzymes break down the molecule to the building blocks. These diseases generally affect paediatric age group.

Main diseases included in sphingolipidoses are Niemann–Pick disease, Tay-Sach's disease, Gaucher disease, Fabry disease, Metachromatic leukodystrophy, Farber's disease and Sandoff's disease. Patient often suffers psychomotor retardation, weakness and spasticity. Antenatal diagnosis is also available (Figs. 14.1 and 14.2).

Sphingolipids catabolism

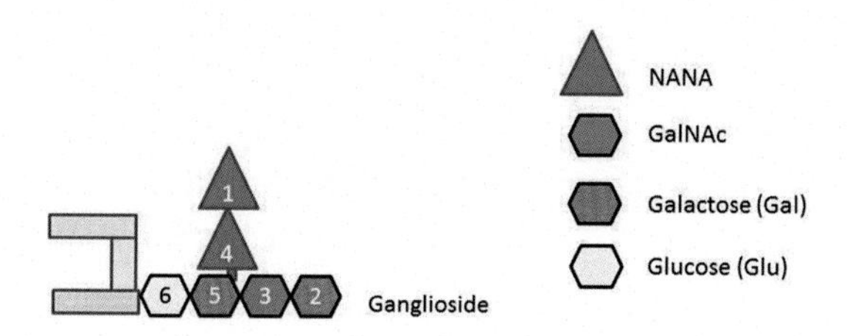

Sweet Biochemistry. DOI: https://doi.org/10.1016/B978-0-443-15348-8.00013-2

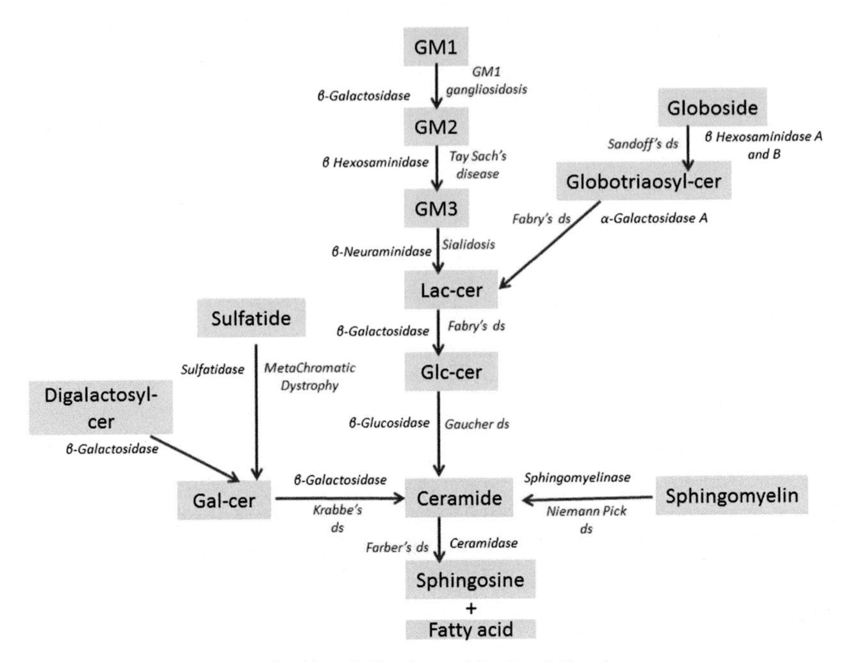

FIGURE 14.1 Catabolism of sphingolipids along with related disorders.

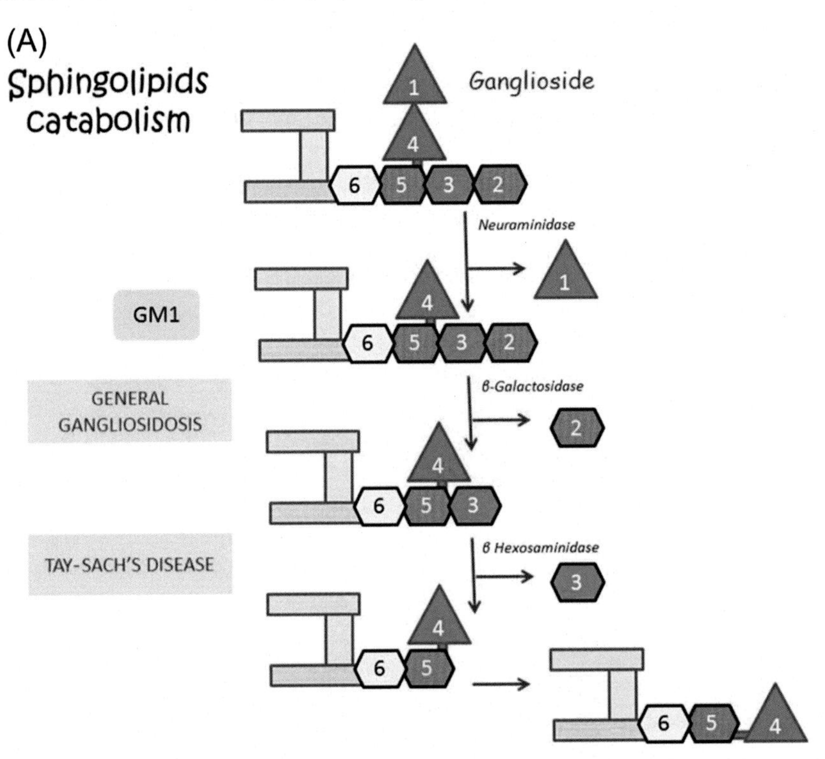

FIGURE 14.2 (A) Steps of sphingolipid catabolism. Observe here the sequence of removal of moieties and the enzymes involved. (B) Sphingolipid catabolism continued and (C) steps of sulfatides and sphingomyelin catabolism.

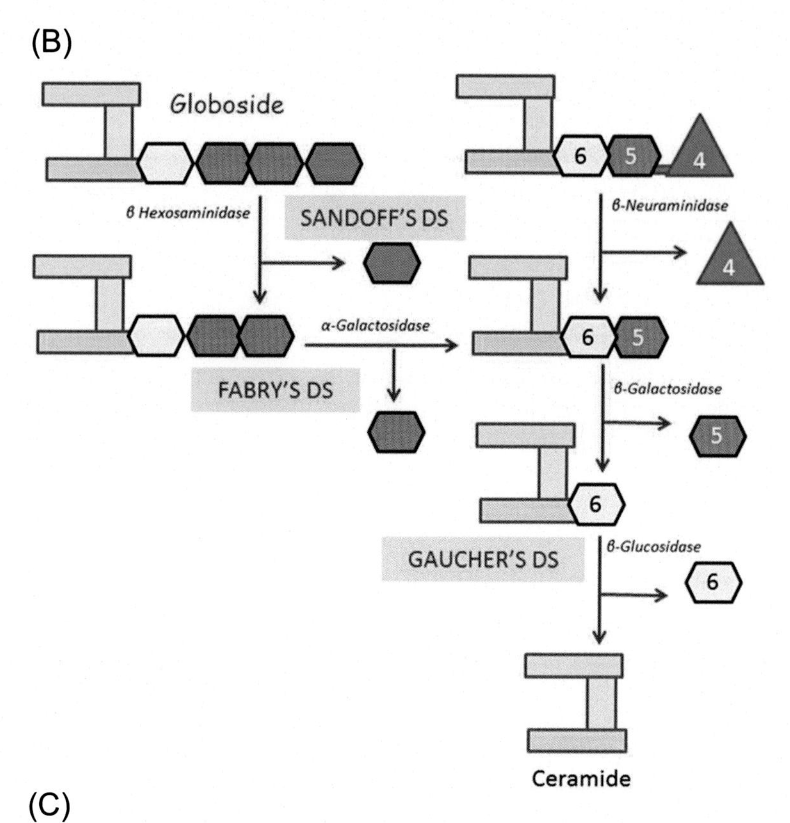

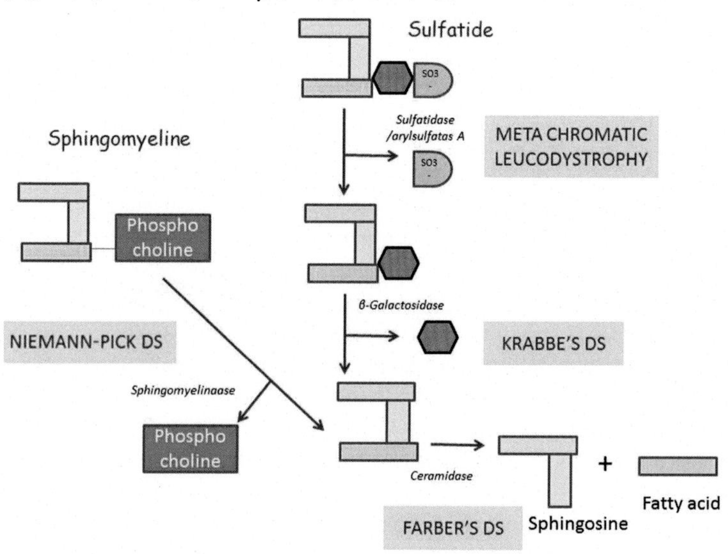

FIGURE 14.2 (Continued).

You need to remember the sequence of removal of components to memorise the degradation of gangliosides to correlate associated diseases. Enzymes' names are also related to substrate e.g., To remove:

NANA – neuraminidase enzyme is used.
GalNAc – hexosaminidase enzyme.
Gal – galactosidase enzyme.
Glu – glucosidase enzyme.

All carbohydrates are attached by beta linkage except first galactose of globoside, sulfatides, globosides and sphingomyelin that are catabolised by sequentially removing components from outer ends as shown in following diagrams.

Tay-Sach's disease and Sandoff's disease

GM_2 gangliosidoses disorders result from a deficiency of the enzyme beta-hexosaminidase. GM_2 Ganglioside is deposited. GM2 disorders include:

A) Tay-Sachs disease (also known as GM2 gangliosidosis-variant B).

It is caused by a deficiency in the enzyme hexosaminidase A. This heterodimer enzyme is constituted by alpha subunit coded by *HEXA* gene on 15th chromosome and beta subunit by *HEXB* gene. Both subunits are capable of cleaving GalNAc residues, but it is the alpha subunit which is capable of break down of GM2 gangliosides. This speciality of alpha subunit is due to Arg-424 residue that binds to N-acetyl-neuraminic residue of GM2 gangliosides. Most common mutation implicated in around 80% of Tay-Sach's patients is a 4-bp addition (TATC) in exon 11 of the HEXA gene. This addition leads to an early stop codon formation.

Symptoms begin by 6 months of age and include progressive loss of mental ability, dementia, deafness, difficulty in swallowing, blindness, cherry-red spots in the retinas and some muscular weakness. Infant starts losing skills after 8–10 months. Children suffering from Tay-Sach's disease usually die from pneumonia and aspiration at less than 6 year age.

Seizures may begin in the second year of life. Due to inability to remove GalNAc from GM2 ganglioside, 100–1000 times more GM2 ganglioside is stored in brain than in the normal person. Anticonvulsant medications are helpful initially Enzyme is detected in tears.

B) Sandhoff disease (variant AB). This gene produces a protein which gives instructions for synthesis of two important enzymes – hexosaminidase A and B present in lysosomes. Onset usually occurs at 6 months of age.

Manifestations of Sandoff disease resemble Tay-Sach's disease but are faster in onset and are of increased intensity (Fig. 14.3).

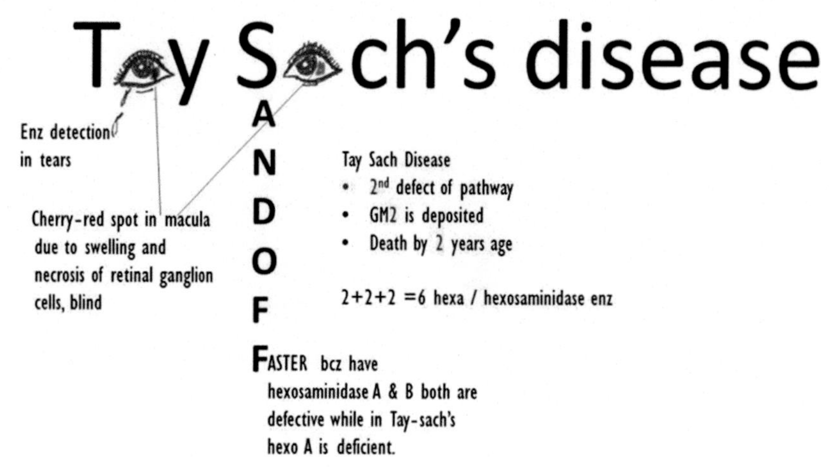

FIGURE 14.3 Tay-Sach's disease.

Gaucher disease

It is a rare inherited deficiency of the enzyme glucocerebrosidase, due to which glucosylceramide is deposited in various organs like bone marrow, spleen and liver. Symptoms like enlarged spleen and liver, liver malfunction, bone lesions causing pain and fractures, severe neurologic complications, swelling of lymph nodes, distended abdomen, a brownish tint to the skin, anaemia, low blood platelets, and yellow spots in the eyes. Gaucher disease type 1, the most common form, is non-neuronopathic because it does not involve the central nervous system. It is associated with thrombocytopenia, anaemia and hepatosplenomegaly. Patients may also suffer from infarction in bones which is highly painful, bone necrosis and osteoporosis. Gaucher disease type 2 is accompanied by neurological complications because of glucocerebroside accumulation in brain. Children experience lose of skills previously acquired along with hypotonia, spasticity, dysphagia, strabismus and failure to thrive. It often leads to fatal aspiration pneumonia. In Gaucher type 3, disease neurological symptoms progress slowly. In less than 5% patients having perinatal-lethal form of Gaucher disease, death occurs before three months age. There is cardiovascular form also (Fig. 14.4).

FIGURE 14.4 Gaucher's disease.

Susceptibility to infection may increase. The disease affects males and females equally, and the inheritance is autosomal recessive. Diagnosis of Gaucher disease is suspected in patient with unexplained anaemia, easy bruising, hepatosplenomegaly and bone fracture. Enzyme assay beta-glucosidase leucocyte is the standard test. Genetic testing for mutations in GBA gene is performed in blood or saliva. Enzyme replacement has been beneficial in the management of patients.

Niemann–Pick disease

This autosomal recessive disorder affects multiple organs in different ranges. It is categorised on the basis of genetic cause and features into type A, type B, type C1 and type C2. Patients with type A disease present with hepatosplenomegaly, failure to thrive, psychomotor regression, interstitial lung disease and cherry-red spot in eye retina. These patients do not cross early childhood. Type B is milder form and has additional features like thrombocytopenia and delayed bone age. This causes short stature and poor bone mineralisation. Type C1 and C2 are different due to genetic basis. Ataxia, vertical supranuclear gaze palsy and poor muscle tone are also observed in these types (Fig. 14.5).

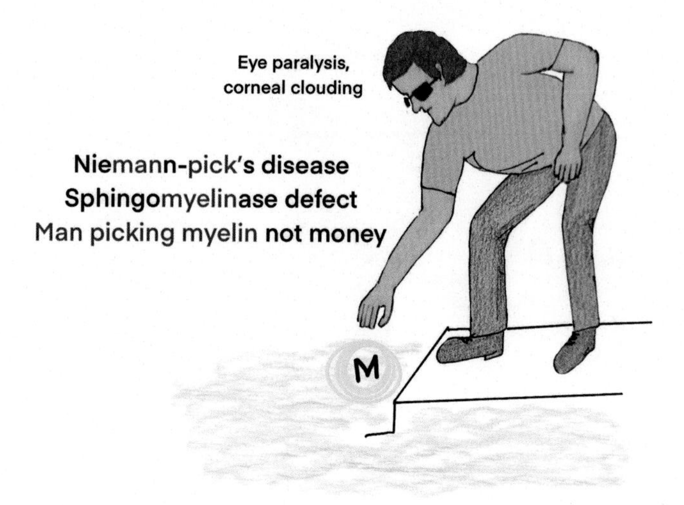

FIGURE 14.5 Niemann–Pick disease.

Sphingomyelin accumulated in cells of the liver, spleen, bone marrow, lungs, and, in some patients, brain. A characteristic cherry-red halo develops around the centre of the retina in 50% of patients. Niemann–Pick disease types A and B are due to *SMPD1* gene mutations encoding sphingomyelinase enzyme. This enzyme breaks sphingomyelin into ceramide. Type C is caused by mutations in genes – *NPC1* or *NPC2* which produce proteins, participating in lipid movement inside the cells.

Farber's disease

Farber's disease, also known as Farber's lipogranulomatosis, is an inherited lipid storage disease. It is due to the deficiency of the enzyme called ceramidase. Mutations in gene ASAH1 code the ceramidase present in lysosomes. Ceramidase breaks ceramide into sphingosine and fatty acid. This autosomal recessive disorder affects both males and females. Ceramide acculumated in the joints, tissues and central nervous system (Fig. 14.6).

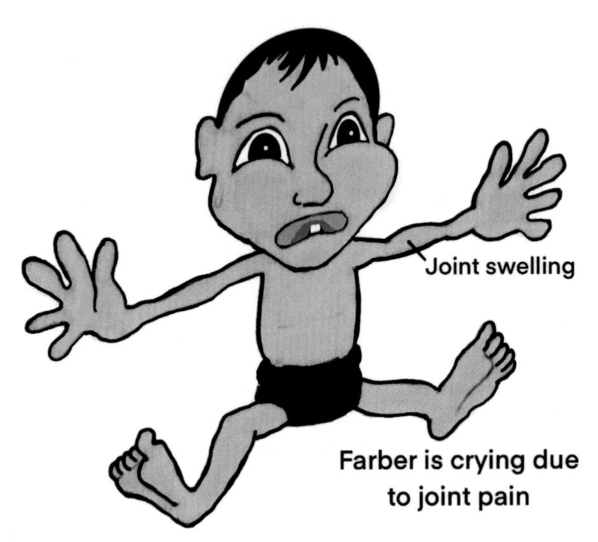

FIGURE 14.6 Farber's disease.

Disease onset is typically in early infancy. Farber lipogranulomatosis is characterised by three features – first a hoarse voice or weak cry, second small fat nodules or lumps below the skin and tissues and third swollen painful joints. Developmental delay, seizures and behavioural problems can affect patients. Moderately impaired mental ability and problems with swallowing are also observed. Death usually occurs by age of 2 years. Hepatosplenomegaly at birth, quadriplegia, myoclonus and hydrocephalus is noticed in severe form.

Fabry's disease

Deficiency of enzyme alpha-glucosidase A is the cause of Fabry's disease. Mutation in gene *GLA* leads to enzyme defect. This is exceptional X-linked lipid storage disease. Due to enzyme deficiency, globotriosylceramide gets accumulated in autonomic nervous system, eyes, kidney and cardiovascular system. Patients experience neurological symptoms like burning pain in arms and legs (acroparesthesias). Cardiomegaly, heart attack and stroke affect the individuals. Progressive renal involvement results in renal failure. Pain abdomen is a common gastrointestinal complaint of patients. On skin, angiokeratomas are noticed which are small, raised, reddish-purple, non-cancerous spots on skin. Decreased sweating, corneal opacity and ringing in the ears are additional features (Fig. 14.7).

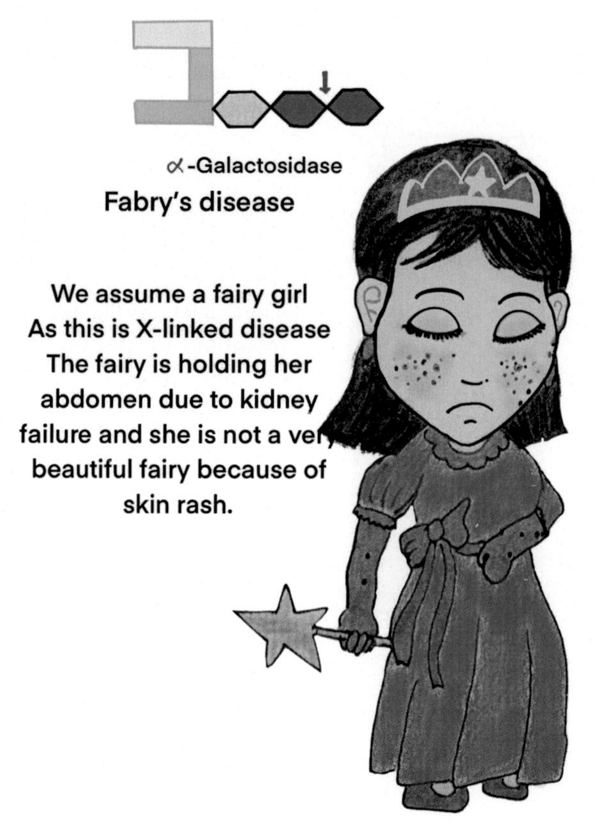

FIGURE 14.7 Fabry's disease.

GM1 gangliosidosis: GM1 gangliosidosis is a genetic disorder in which progressive accumulation of GM1 ganglioside in nerve cells destroys the neurons in brain and spinal cord leading to neurodegeneration and seizures. The deposition is due to deficiency of lysosomal enzyme beta-galactosidase which breaks down GM1 ganglioside required for normal functioning of brain. The genetic cause of deficiency is mutation in *GLB1* gene. In other organs, the deposition causes hepatosplenomegaly, coarsening of facial features, skeletal irregularities, muscle weakness, joint stiffness, cherry-red spot in retina and angiokeratomas (Fig. 14.8).

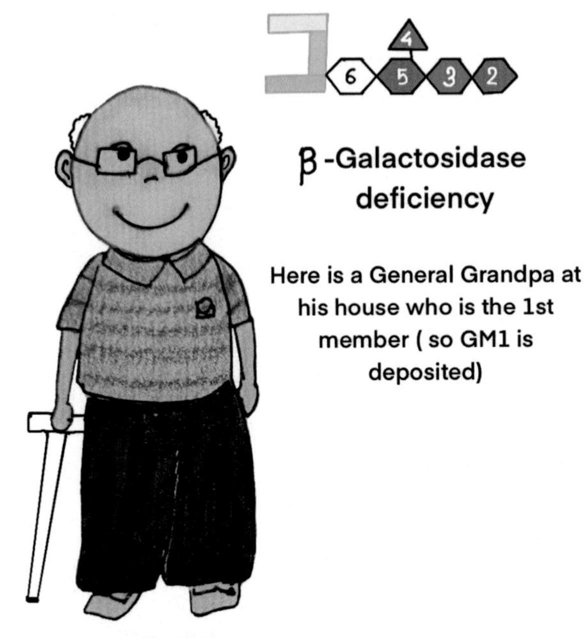

FIGURE 14.8 General gangliosidosis.

Krabbe disease: Krabbe disease is a rare autosomal recessive genetic disorder also known as globoid cell leukodystrophy and galactosylceramide lipidosis. In this disease, enzyme galactocerebrosidase defect due to mutation in *GALC* gene leads to piling up of galactosylceramide and psychosine in various cells and tissues including brain cells. The characteristic feature is globoid cells which are cells having more than one nucleus. These cells damages the myelin sheath of neurons resulting in rapid deterioration of mental and motor skills. Infants are mostly affected with the onset usually starting before 6 months. Symptoms include feeding problems, unexplained fever, muscle weakness, hypertonia, myoclonus seizures, deafness, spasticity and blindness. Child seldom passes age beyond 2 years (Figs. 14.9 and Fig. 14.10).

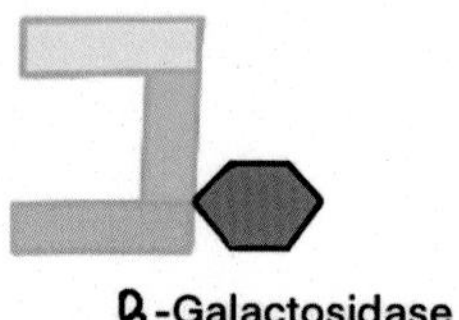

FIGURE 14.9 Krabbe disease.

FIGURE 14.10 Three Galactosidase deficients standing together. A is Fabry's disease with alpha-Galactosidase defect. B is GM1 Glycosidosis and C is Krabbe's disease. Both of these later ones are beta-Galactosidase defects.

Metachromatic leukodystrophy

Metachromatic leukodystrophy or MLD is a lipid storage disease characterised by deposition of fats called 3-sulfogalactosylceramide in the white matter of the central nervous system, peripheral nerves and kidneys. The name indicates the view under microscope where sulfatides collected in cells resemble granules that are coloured differently from the rest of cellular material (Fig. 14.11).

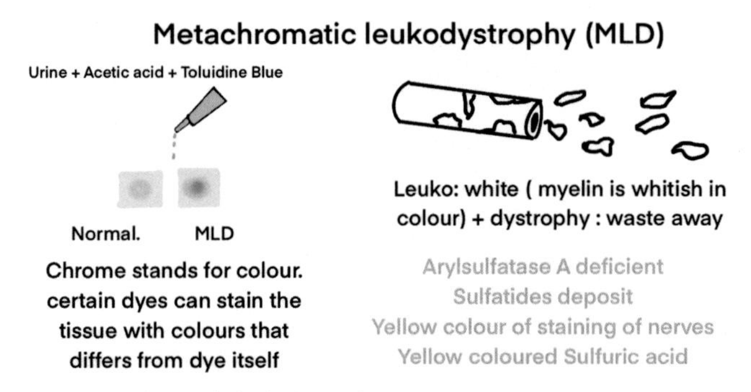

FIGURE 14.11 Metachromatic leukodystrophy.

Myelin sheath is disrupted in brain. Autosomal recessive disorder is caused by mutations in *ARSA* gene and sometimes *PSAP* gene. These mutations lead to deficiency of the enzyme arylsulfatase A.

Onset typically starts between age of 12 and 20 months. Infants may appear normal at first but develop difficulty in walking and a tendency to fall, followed by intermittent pain in the arms and legs, progressive loss of vision leading to blindness, developmental delays, impaired swallowing, convulsions, and dementia before the age of 2 years. Death usually occurs before 5 years.

Chapter 15

Ceramide structure and derivatives

Ceramide belongs to waxy lipid biomolecules family (Latin Cera word means wax). Structure of ceramide is constituted by sphingosine alcohol and a fatty acid linked by an amide bond. Ceramide is precursor for many sphingolipids which are ubiquitously present in eukaryotic cell membrane, involved in myriad signalling pathways (Fig. 15.1). The length of fatty acid may vary from C14 to C25. Ceramide becomes influential due to their presence in sphingomyelin present in myelin sheath of neurons. Ceramide and their derivatives are endowed with functions like regulating differentiation, cell proliferation and apoptosis.

Ceramide and Sphingosine structure

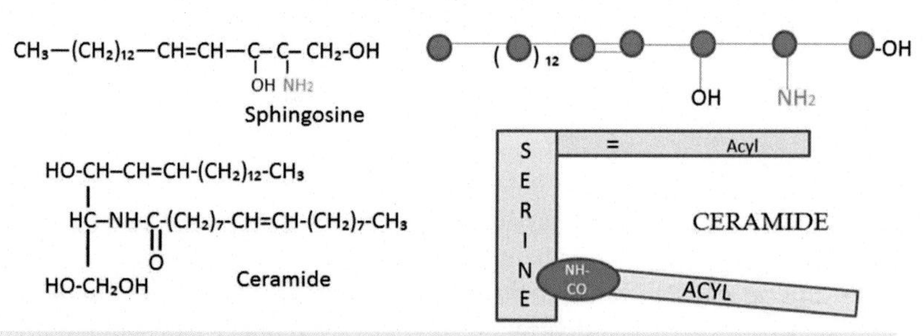

Sphingosine is a 18-carbon amino alcohol with a long unsaturated hydrocarbon chain.

Ceramide is a long-chain fatty acid amide derivative of sphingosine or simply it is the combination of sphingosine alcohol with fatty acid. Derivatives of ceramide are formed by attaching to OH group on C1.

Ceramide plays an important role in signal transduction in programmed cell death **(apoptosis),** the cell cycle and cell differentiation and senescence. Ceramide is the core molecule of sphingolipids.

FIGURE 15.1 Ceramide and sphingosine structure.

Sweet Biochemistry. DOI: https://doi.org/10.1016/B978-0-443-15348-8.00001-6

The sphingosine is a nitrogen-containing acyl amino alcohol synthesised from palmitoyl-CoA and serine. The acyl group is attached at amino group. This rate-limiting reaction requires pyridoxal phosphate (PLP), nicotinamide adenine dinucleotide phosphate (NADPH) and Mn++. 3-keto sphinganine formed here is reduced to dihydro sphingosine (Fig. 15.2).

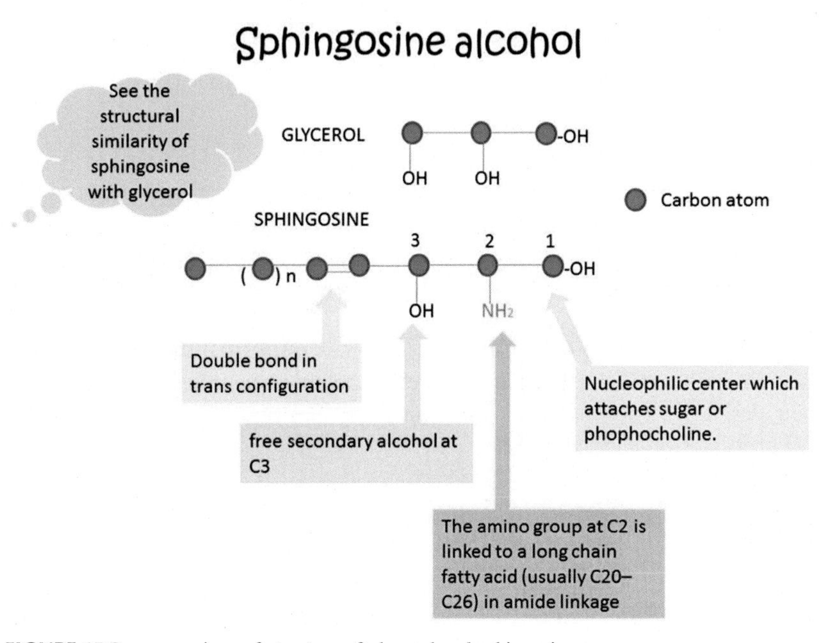

FIGURE 15.2 comparison of structure of glycerol and sphingosine.

The sources for fatty acid component fatty acid synthase complex and microsomal chain elongation system. Oleic fatty acid, nervonic acid and alpha hydroxylated very long-chain fatty acids can also be attached.

Substitution of molecules is possible at terminal C1 hydroxyl group of ceramide. So two branches diverge – sphingomyelin and glycosphingolipids

(GSLs). In the former molecule, a sphingosine base with 18-carbon chain and a double bond at position four joins with phosphorylcholine fatty acid. Hence, they can be called phospholipids also (Fig. 15.3).

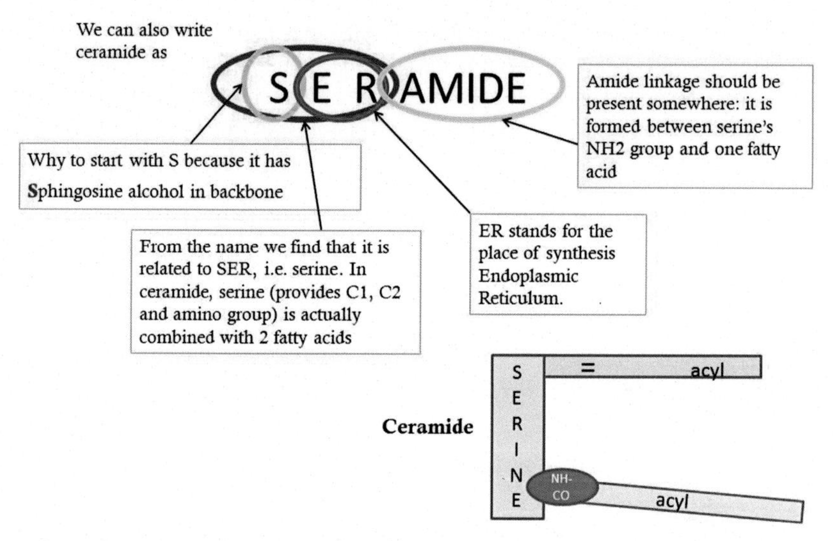

FIGURE 15.3 Ceramide structure learning.

GSLs or glycolipids are also formed by union of carbohydrates with ceramide.

In ganglioside, one of the sugars is acidic in nature, like N-acetylneuraminate or N-glycolylneuraminate. These sugars have been given the name 'sialic acids'. GSLs may be neutral (due to neutral sugar without any charge), sialylated (with sialic acid residues) or sulphated. Second-type GSLs are widely known as 'Ganglioside'.

Structure wise, ceramides are amphipathic molecules like cholesterol. This is due to a polar head and a non-polar fatty acid. The special structure makes ceramide one of the favourite molecules in membranes. The double bond in sphingoid base promotes intramolecular H-bonding in the polar region. This helps in close packing of ceramides. It is easy to understand that the length of fatty acid chain decides curves in membranes. Ceramides with smaller fatty acids lead to positive curvature, while long-chain ceramides have a high intrinsic negative curvature. Isn't this interesting?

What are ceramides doing in our body

Ceramide's interaction with ion channels modulate the membrane permeability for solutes. Ceramides collect as rafts or microdomains in membrane which actively involved in signal transduction. The process of exosome formation and secretion requires changes in membrane curvature, dependent on ceramides.

Approximately half of lipid content in bilayer is constituted by ceramides in stratum corneum layer of skin. Fatty acids of ceramides are mostly aliphatic, non-branched, saturated long chains having high melting point. At physiological temperatures, the lateral diffusion exhibited is minimal. No wonder why ceramides being the essential membrane component are added to the cosmetics nowadays.

Some ceramide derivatives are proapoptotic (Fig. 15.4). Intracellular caspases are released from the membrane by channel formation. Thus, ceramides have been implicated seriously in apoptosis. Recent research is unveiling the role of ceramides in autophagy also. Because of these two roles, ceramides are significant for tumour suppression and any defects in the pathways for ceramides may allow cancer cells to escape.

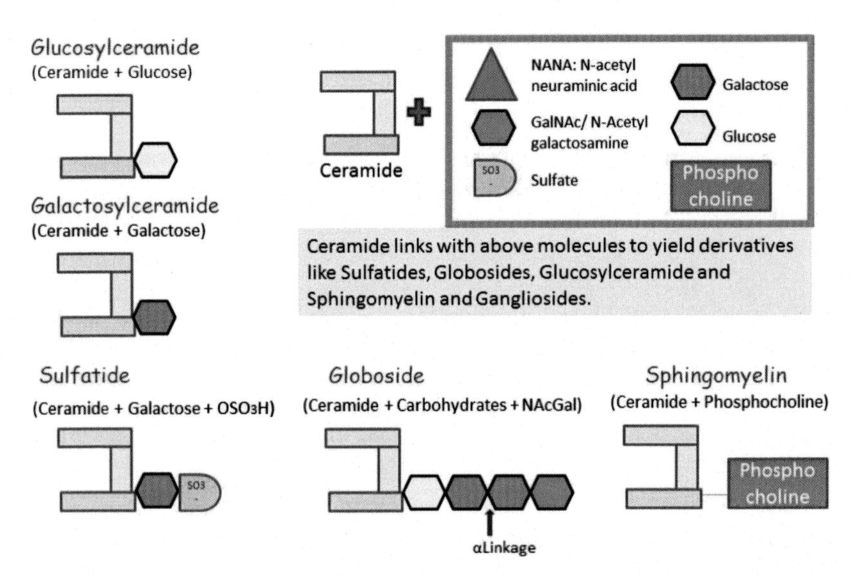

FIGURE 15.4 Ceramide derivatives composition.

Ceramides participate in enzyme regulation by activating protein phosphatases and protein kinases. Overall, ceramides promote catabolism and slow down anabolism (Figs. 15.5 and 15.6).

Ceramide Derivatives

Galactocerebroside and **Glucocerebroside** are ceramide linked with galactose or glucose. These carbohydrate are joined with the help of activated nucleotide sugars UDPgalactose and UDPglucose.

Sulfatide or sulfogalactocerebroside, is a sulfuric acid ester of galactocerebroside.

Globosides are cerebrosides having two or more sugar residues, commonly galactose, glucose, or *N*acetylgalactosamine (GalNAc)

Sphingomyelin, is ceramide phosphocholine and **an important component of myelin sheath of neurons.** Hydroxyl (-OH) group at C1 Of sphingosine is esterified to choline through a phosphodiester bond and the NH_2 group unites with a long chain fatty acid through a amide bond to form sphingomyelin. Thus sphingomyelin is also a phospholipid. It is neutral at physiological PH because of equal positive and negative charge.

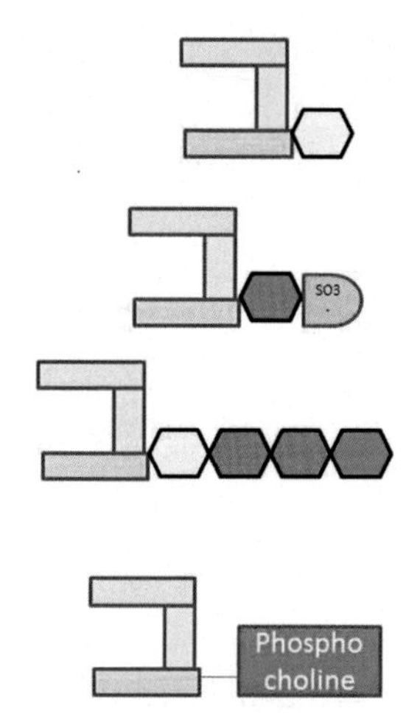

FIGURE 15.5 Ceramide derivatives.

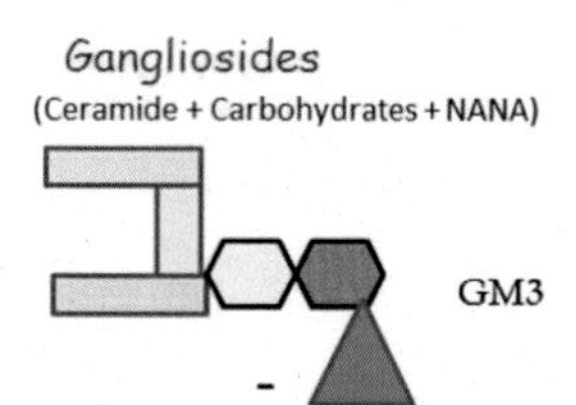

Ganglioside are sialic acid rich glycosphingolipids abundantly present in ganglion cells of the central nervous system.
Gangliosides are written as GM1, GM2 and GM3 where G indicates ganglioside while M stands for monosialo-containing species. Numbers 1,2 or 3 are assigned according to chromatographic migration. GM1 is the more complex and GM3 is the simplest ganglioside.

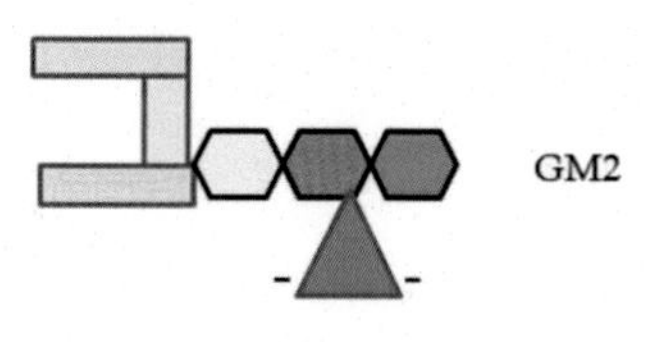

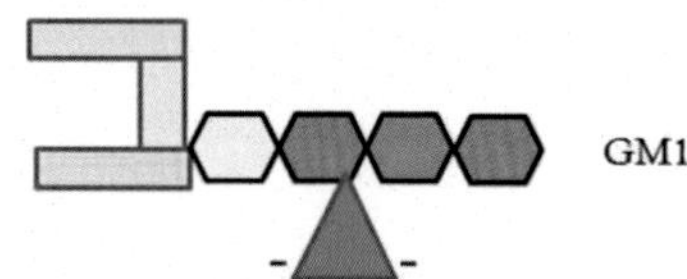

FIGURE 15.6 Gangliosides.

Chapter 16

Prostaglandin synthesis

Prostaglandins (PGs) are physiologically and pharmacologically active, lipid compounds resembling hormones which act through G-protein-linked receptors. PGs are found in almost all nucleated cells of body, but the name PG was coined due to their discovery in prostate gland. They elicit their biological hormone-like actions locally as they are short-lived. The effect of PGs varies in different cells because of binding to different receptors. Principal actions of PGs include inflammation stimulation, regulation of blood flow, and sleep cycle induction, affecting membrane transport and modulating synaptic transmission in nervous system.

PGs are derived from essential fatty acids of membrane. Chemically, PGs are C20 saturated fatty acid derivatives having a 5-carbon ring. PGs belong to eicosanoids family (Greek eikosi means 20) which arise from arachidonic acid. Other members of eicosanoids are thromboxanes (TX), leukotrienes (LT) and lipoxins (LX). Arachidonic acid (AA) can be a substrate for two enzymes: cyclooxygenase (COX) and lipooxygenase (LOX). When AA is a substrate for COX, PG2 and TX2 series are synthesised, while on entering LOX pathway, LT4 and LX4 series are formed. According to the ring present in a PG, it is given a letter which is followed by a number indicating the number of double bonds, for example, PGE1, PGE2, PGI2, and PGF2α. Alpha denotes the OH projection on carbon 9 in cyclopentane in PG.

Arachidonic acid, a 20:4 fatty acid derived from linoleate essential fatty acid, undergoes sequential oxidations and isomerisations, to form PG. In PG synthesis, arachidonic acid is converted to PGG2 by the action of COX enzyme. PGG2 is oxidised to PGH2 which can be further isomerised to PGD2, PGE2 and PGF2α.

Prostaglandin E2 (PGE2) is generated from the action of prostaglandin E synthases on prostaglandin H2 (prostaglandin H2, PGH2).

Alternatively, PGH2 can be a substrate for thromboxane synthase to yield TXA2 and TXB2. PGH2 can also form prostacyclins in the presence of prostacyclin synthase (Fig. 16.1).

Sweet Biochemistry. DOI: https://doi.org/10.1016/B978-0-443-15348-8.00030-2

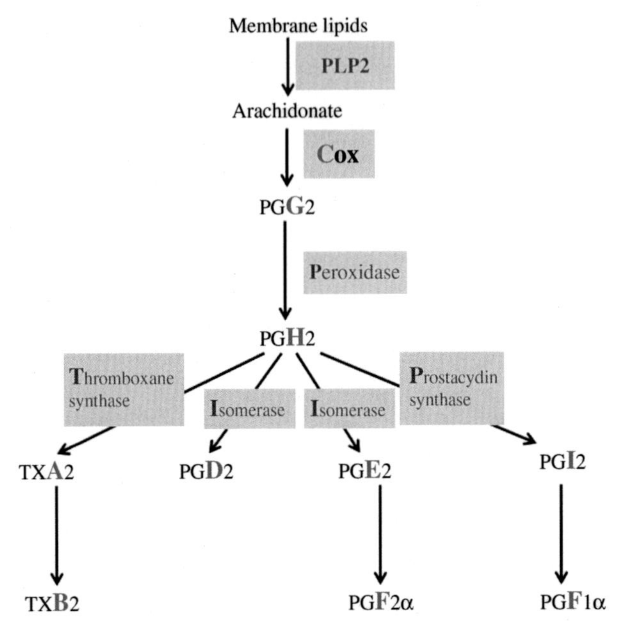

FIGURE 16.1 Prostaglandin synthesis.

Biological actions of prostaglandins

Cardiovascular system: PGI2 – vasodilation inhibits platelet aggregation, TXA2 – vasoconstriction inhibits platelet aggregation.

Ovary and uterus: PGE2 – uterine muscle contraction.

Respiratory system: PGF – bronchoconstriction, PGE – vasodilator.

Immune system and inflammation: PGE2 and PGD2 – proinflammatory and increase capillary permeability.

Gastrointestinal tract: inhibits gastric secretion and increase motility of gut.

Metabolism: PGE2 increases glycogen synthesis, decreases lipolysis and mobilises calcium from bones.

PG synthesis is catalysed by microsomal enzymes in a stepwise manner (Fig. 16.2).

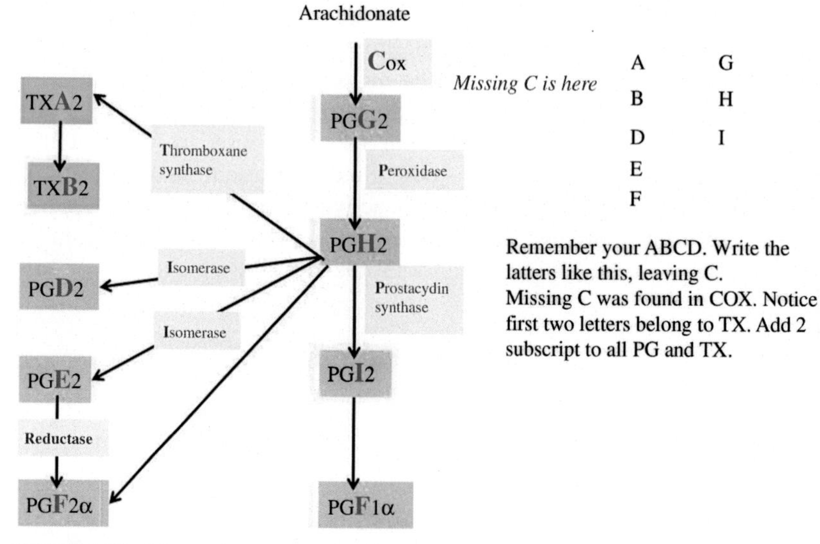

FIGURE 16.2 Mnemonic for prostaglandin synthesis.

Cyclooxygenase is the controller of rate-limiting step

COX enzymes are tyrosyl-radical using homodimer hemoproteins possessing two activities: dioxygenase and peroxidase activity. COX enzyme is also known as prostaglandin H synthase. There are two isoforms of COX: 1 and 2. The amino acid composition of two forms is 60% identical, and the homology at active site is almost 90%. In both forms, there is a similar long narrow L-shaped hydrophobic channel with a hairpin bend at one end to let the substrate enter. It is interesting to note that in this channel at position 523, the amino acid is isoleucine in COX-1 and valine in COX-2. This single amino acid difference allowed synthesis of COX-2 inhibitors.

There are three component domains of each monomer of COX: the epidermal growth factor domain, the membrane binding helical domain and the catalytic domain having COX and peroxidase activities on either sides of prosthetic group haem.

COX enzyme inserts two oxygen molecule (O_2) in the arachidonic acid structure. But this does not happen in single step. The process starts with the abstraction of hydrogen on carbon 13 by COX and insertion of 1 O_2 on C11. The special feature of this step is the antarafacial insertion. Don't be confused with the terms. Antarafacial insertion simply means that the plane of oxygen insertion is opposite from that of hydrogen abstraction. Abstraction is a bimolecular chemical reaction in which an atom that is either neutral or charged is removed from a molecular entity. In this way, a double bond appears between C12 and C13 along with removal of double bond between C8 and C9. Also one oxygen atom newly added forms a link with C9. A cyclopentane ring is formed in arachidonic acid by relocalisation of electrons.

Then the addition of second oxygen molecule (O_2) takes place at C15. The inserted oxygen is in the form of hydroperoxide (−OOH) group, and the product molecule is called prostaglandin G2. Reduction by COX at this hydroperoxide results in hydroxyl (−OH) group formation. At this stage, the prostaglandin H2 is formed by peroxidase activity of COX enzyme.

It is important to discuss the differences between two forms due to functions, regulation and targeted treatments (Table 16.1).

TABLE 16.1 Difference between COX-1 and COX-2 (*COX*, cyclooxygenase).

	Feature	COX-1	COX-2
1	Amino acids number	576	581
2	Specificity	More specific for arachadonic acid over linoleic acid due to smaller active site	Less specific due to larger active site
3	Mannose oligosaccharide number	Three	Four. One extra is for the regulation of degradation
4	Expression	Constitutive	Inducible, normally undetectable
5	Related gene	Gene for COX-1 has a promoter region without a TATA sequence	Gene for COX-2 has sequences for rapid upregulation in response to stimuli
6	Functions	Gastric cytoprotection, decreasing acid secretion in stomach, regulating renal blood flow, haemostasis and homoeostasis in vascular system	Pain, mediation of inflammation, cell differentiation, mitogenesis, maturation and normal development of kidney and brain

Substrate specificity. For example, COX-2 is capable of metabolising ester and amide derivatives of AA that are poor substrates for COX-1.

COX inhibitors were named nonsteroidal antiinflammatory drugs, the representative being aspirin (acetylsalicylate). Aspirin is commonly used antiinflammatory and analgesic over-the-counter drug. It irreversibly inhibits both COX forms by the acetylation of hydroxyl group of Serine-530 residue present at the active site of enzyme.

Some organic acids are reversible inhibitors of COX, and mechanism is the formation of hydrogen bond with arginine residue at position 120. Examples of such drugs are ibuprofen and indomethacin.

Drugs which blocked both COX-1 and 2 caused many unwanted side effects; thus, the need of selective COX-2 inhibitors arose. This group decreased the incidence of gastrointestinal injury in patients. Usually, Cox-2 inhibitors had sulfonyl, sulphone or sulphonamide group that can bind COX-2 side pocket in the hydrophobic channel. Celecoxib and Rofecoxib belong to this family.

Chapter 17

Purine structure

A purine is a naturally occurring, aromatic heterocyclic water-soluble molecule composed of carbon and nitrogen. Purines include adenine and guanine which participate in DNA and RNA formation. Purines are also constituents of other important biomolecules, such as adenosine triphosphate (ATP), guanosine -5′- phosphate (GTP), S-adenosyl methionine, cyclic adenosine mono phosphate (cAMP), nicotinamide adenine dinucleotide (NADH) and coenzyme A. Examples of proteins associating with purines for their activities are nucleic acid polymerases, kinases, nucleotide phosphorylase and few receptors. Purine-rich dietary sources are sweetbreads, liver, beef, kidney, brain, meat extracts, yeast and Herring, while moderate amounts are present in red meat, fish, asparagus, cauliflower, spinach and lentils, etc.

Purines are weakly acidic (pKa 8.93) as well as weakly basic (pKa 2.39) by molecular nature. By the presence of NH_2 group and oxo groups, purines exhibit keto-enol and amine-imine tautomerism, although amino and oxo forms predominate in physiological conditions. Substituents additions and interaction with other biomolecules can modulate tautomeric forms.

Basic purine has nine atoms in its structure. Purine has two cycles: six-membered pyrimidine ring and five-membered imidazole ring fused together. Four nitrogen atoms are present at 1, 3, 7 and 9 position. Numbering of purine starts with first nitrogen of six-membered ring and then proceeds in anticlockwise direction. Imidazole ring is numbered clockwise. Other important purines include hypoxanthine, xanthine, theobromine, caffeine, uric acid and isoguanine (Fig. 17.1). Purine bases connect with Carbon-1' of pentoses through ninth nitrogen atom to form nucleosides.

Sweet Biochemistry. DOI: https://doi.org/10.1016/B978-0-443-15348-8.00022-3

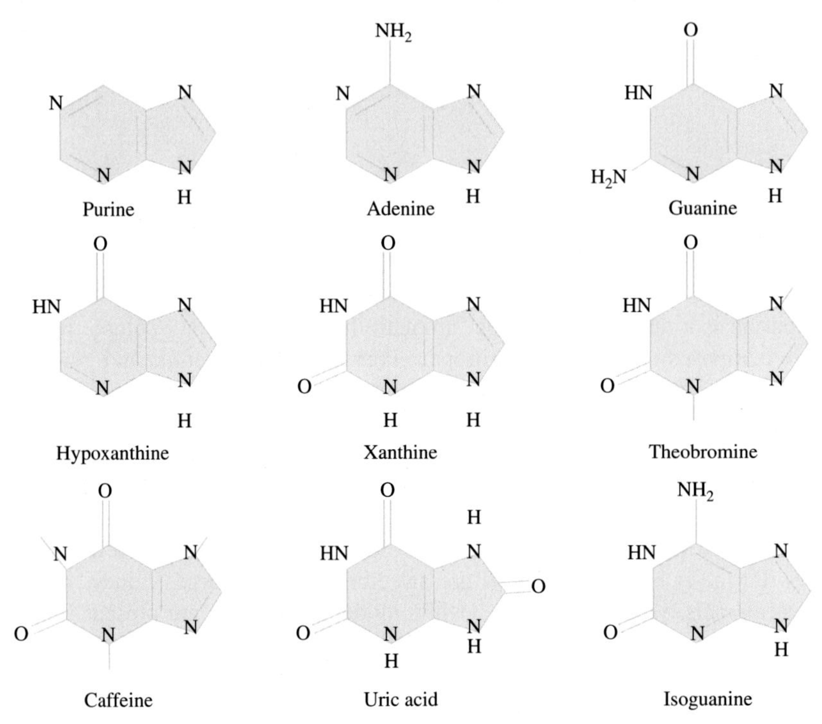

FIGURE 17.1 Structures of various purines.

Source of atoms of purine bases is as follows: (Fig. 17.2)

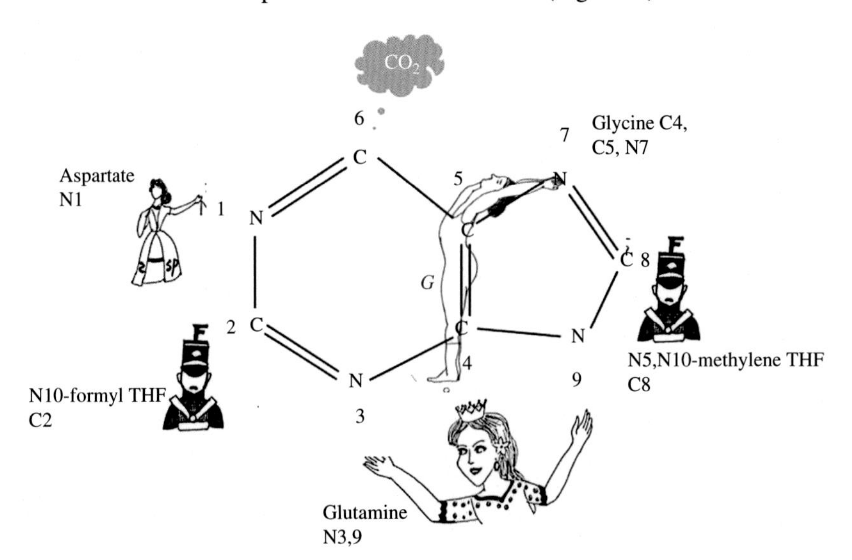

FIGURE 17.2 Picture mnemonic for sources of atoms of purine.

Nitrogen 1 – amino group of aspartate.
Carbon 2 – formyl THFA.
Nitrogen 3 – amide N of glutamine.
Carbon 4, carbon 5, nitrogen 7 – glycine.
Carbon 6 – respiratory CO_2.
Carbon 8 – methylene THFA.
Nitrogen 9 – amide N of glutamine.

Chemical structure

Adenine = 6-amino purine, Guanine = 2-amino, 6-oxypurine, Hypoxanthine = 6-oxy purine, Xanthine = 2,6-dioxy purine, Uric acid = 2,6,8-tri-oxopurine.

Different sources of atoms of purine are depicted here. For Nitrogen 1, a lady wearing dress with Asp written on it is drawn. Carbon 2 and carbon 8 are donated from formyl THFA and methylene THFA soldier. Do notice their cap with a big F for folate. Nitrogen 3 and 9 are gifted by princess glutamine. Carbon 4, 5 and Nitrogen 7 are embedding a glycine in structure. Lastly, the carbon 6 is taken from clouds of carbon dioxide.

Examples of Interesting purines

Caffeine: An water-soluble alkaloid with structure 1,3,7-trimethylxanthine formed by fusion of six-membered pyrimidinedione ring and five-membered imidazole rings. Caffeine is consumed globally in caffeinated drinks and is derived from coffee beans, tea leaves and cacao pods. It reaches a peak concentration within half an hour. Caffeine is known to raise dopamine levels, cause cerebral vasoconstriction and stimulation of central nervous system (CNS), cardiovascular system (CVS) and BP-regulating centres. In skeletal muscles and adipose tissue, caffeine elevates intracellular cyclic AMP by inhibiting enzyme phosphodiesterase. This activated hormone-sensitive lipase and promotes lipolysis causing release of free fatty acids and glycerol. This explains why people use black coffee for weight loss.

Theobromine: Theobromine, alkaloid derivative of *Theobroma cacao*, is also called xantheose. Even with bitter taste, it finds a place in chocolates. It is also present in tea plant and kola nut. Theobromine is less water-soluble than previous alkaloid caffeine. It acts as a bronchodilator and works by inhibiting adenosine receptors.

Theophylline: Another commonly used stimulant is theophylline, methylxanthine deriative from tea. It brings effects like diuresis, smooth muscle relaxation, bronchodilation, and stimulation of CNS and CVS. Most of us have a first-hand experience of effects of tea. Theophylline blocks adenosine blocker and activates histone deacetylase. Theophylline is indicated for patients of asthma, bronchospasm and COPD.

Uric acid: Uric acid is a well-studied molecule due to implication in Gout, a crippling arthritis and nephrolithiasis. It forms ion and salts referred as urates. At physiological pH, this diprotic acid exists as monosodium urate. Uric acid can shuffle between Lactam and Lactim forms (ketl-enol tautomerism). One of the main problems with uric acid is that it's salts are less water-soluble, and this is promoted by low temperature. Moreover, the recrystallisation is very easy. Despite some ill effects, uric acid is major antioxidant present in human plasma.

Chapter 18

Purine synthesis de novo

Purines are synthesised majorly by liver through cytoplasmic multistep de novo synthesis pathway. De novo synthesis means that the rings are made by compiling atoms from their sources. This pathway is tightly regulated. Purine rings are synthesised on platform of ribose-5-phosphate to yield nucleotides. De novo synthesis pathway takes place in 10 steps. Phosphoribosyl pyrophosphate (PRPP) provides ribose-5-phosphate as starting material. De novo synthesis from PRPP to IMP is a collaborative work of six enzymes. These enzymes form a multienzyme complex known as purinosome around mitochondria and microtubules. The enzymes include one trifunctional, two bifunctional and three monofunctional enzymes. One molecule of IMP generation requires five ATP, two glutamine and formate molecules, one glycine, 1 aspartate and 1 carbondioxide. These inputs are crucial in states of rapid cell proliferation like cancer.

A brief view of reactions

In the first reaction which is also the committed step, ammonia (N9 of final purine) released from glutamine displaces pyrophosphate to produce *5-phosphoribosyl-1-amine*. Enzyme governing this step is *Glutamine phosphoribosyl amidotransferase/Amido phosphoribosyl transferase (Atase)*. Following reactions share a similar mechanism in which carbon-bound oxygen atom is activated by phosphorylation, and then the phosphoryl group is displaced by ammonia or an amine group acting as a nucleophile (Figs. 18.1 and 18.2).

Sweet Biochemistry. DOI: https://doi.org/10.1016/B978-0-443-15348-8.00009-0

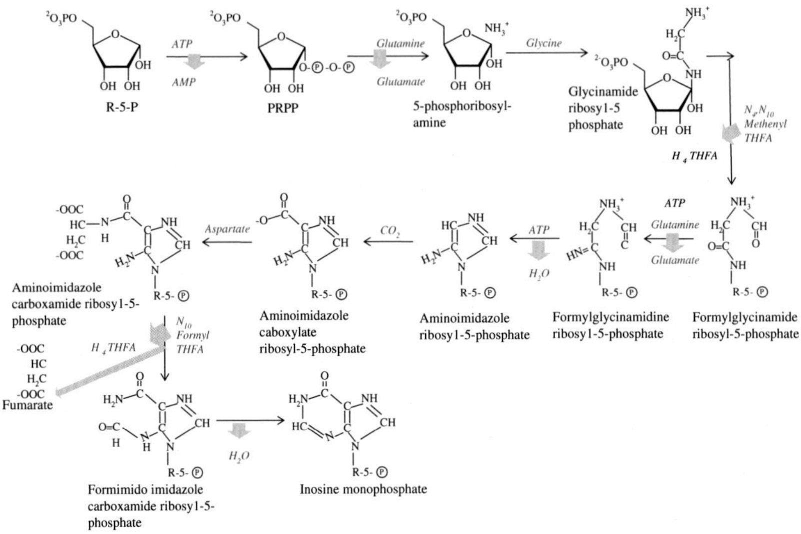

R-5-P
ATP
AMP
PRPP
Glutamine
Glutamate
5-phosphoribosyl-amine
Glycine
Glycinamide ribosyl-5 phosphate
N_5N_{10} Methenyl THFA
H_4 THFA
Formylglycinamide ribosyl-5-phosphate
ATP
Glutamine
Glutamate
Formylglycinamidine ribosyl-5-phosphate
ATP
H_2O
Aminoimidazole ribosyl-5-phosphate
CO_2
Aminoimidazole caboxylate ribosyl-5-phosphate
Aspartate
Aminoimidazole carboxamide ribosyl-5-phosphate
Fumarate
N_{10} Formyl THFA
H_4 THFA
Formimido imidazole carboxamide ribosyl-5-phosphate
H_2O
Inosine monophosphate

FIGURE 18.1 Purine de novo synthesis with molecular structures of intermediates.

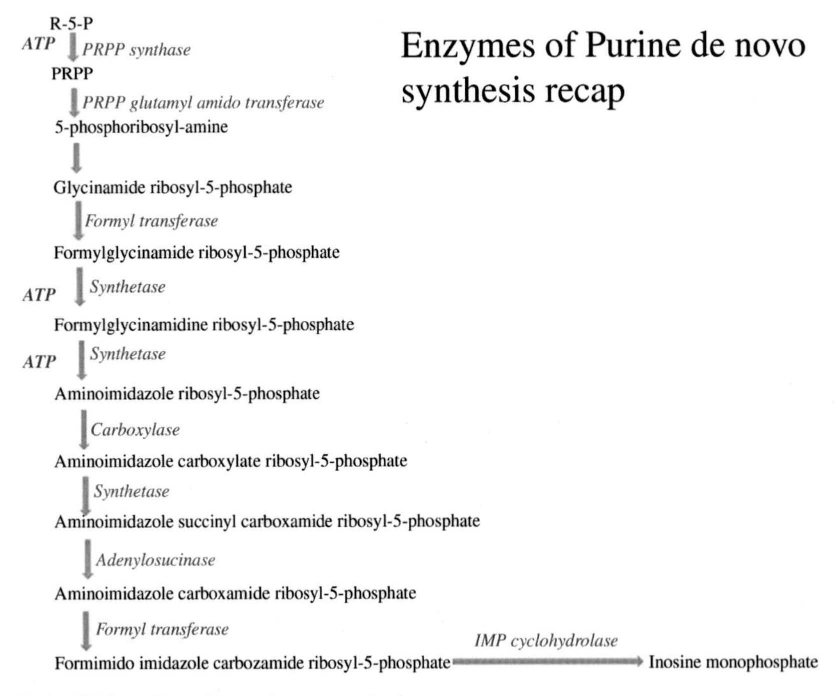

FIGURE 18.2 Basic Purine de novo synthesis.

Glycine molecule is almost completely consumed (C4,C5,N7) in the growing ring by joining the amino group of phosphoribosylamine. In the next step, a formyl group (C8) is transferred to amino group of glycine component by N 10-formyltetrahydrofolate. Before closure of five-membered imidazole ring, amidine group is formed by ammonia released from glutamine. This adds third nitrogen (N3) of purine ring (Figs. 18.3 and 18.4).

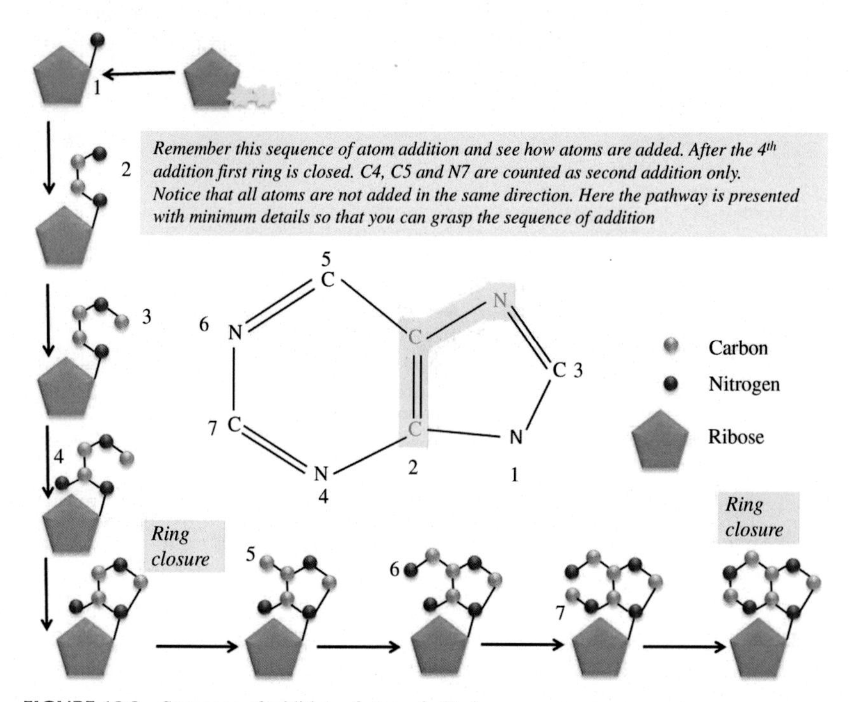

FIGURE 18.3 Sequence of addition of atoms in Purine- an easy way.

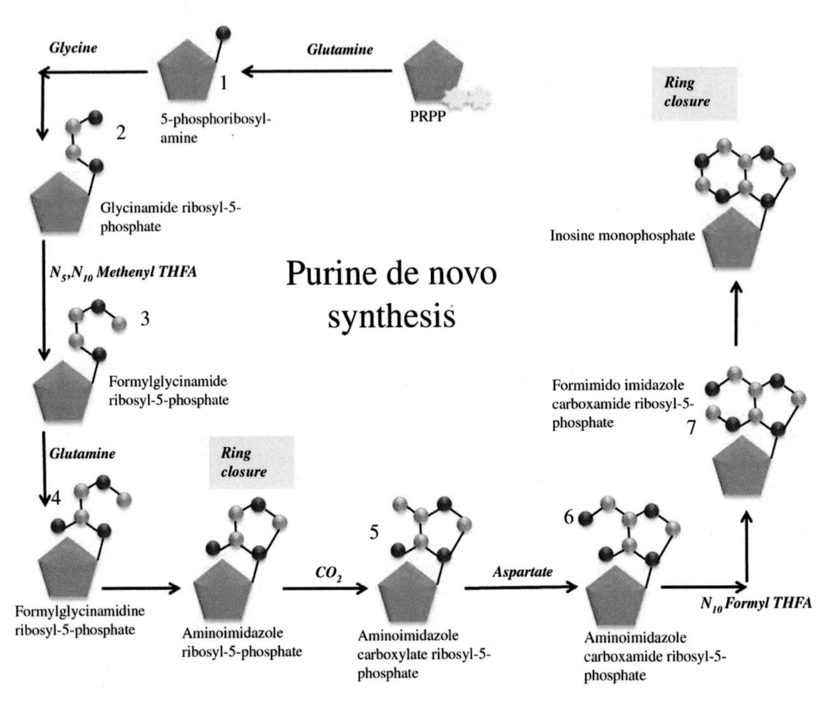

FIGURE 18.4 Correlating the addition of purine atoms with substrates.

Five-membered imidazole ring is closed utilising and an ATP. Carbondioxide in the form of bicarbonate adds C6 to imidazole ring. The imidazole carboxylate is phosphorylated followed by the displacement of the phosphate by the amino group of aspartate. Fumarate is released, leaving behind amino group. Second formyl group (C2) is added from N 10-formyl THFA and second cyclisation yields inosinate (IMP).

I understand we need some trick to learn purine de novo synthesis as this is quite complicated.

Reader must not confuse the numbering of purine atoms with the sequence of atom additions. Numering of purine ring is shown in A where pyrimidine part is numbered anticlockwise and imidazole ring is numbered clockwise (Fig. 18.5).

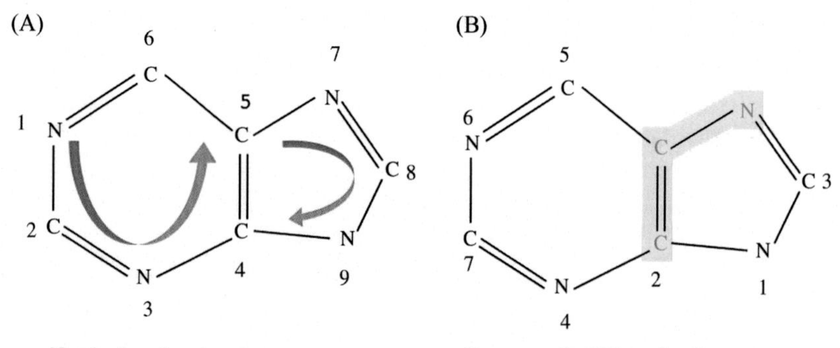

FIGURE 18.5 Differentiating the numbering of purine atoms from the sequence of addition of a purine atom.

Sequence of addition of purine atoms is shown in B which is not in linear direction. Sequence of addition of atoms when used with sources of atoms can indicate the reactions. This makes the intricate pathway simpler.

Amido phosphoribosyl transferase (Atase) – rate-limiting enzyme with a unique ammonia channel

Atase has two important domains – glutaminase domain and a phosphoribosyltransferase domain. The name of domain explains the activity conducted, i.e., glutaminase domains hydrolyses glutamine amino acid to release ammonia. Second domain attaches ammonia to ribose-5-phosphate. Catalytic site of Atase is special to have an oxyanion hole for acting on reaction intermediate. Atase uses ammonia as a nucleophile. The substrate PRPP activates enzyme by bringing a conformational modification in glutamine loop which results in 200-fold higher Km value for binding of glutamine. But the changes continue even after this at active site, making the glutamine inaccessible. These conformation changes are associated with manifestation of an hydrophobic ammonia channel that is 20 Å long. In the channel, hydrogen bonding sites are absent for easy transport of ammonia from glutamine to PRTase catalytic site.

Atase is inhibited by feedback of products like IMP, AMP and GMP, while its substrate PRPP can activate it. As this enzyme catalyses the first step of pathway, when PRPP concentration rises in cell, de novo purine synthesis is enhanced leading to elevated uric acid levels. Such condition is seen in patients with Lesch–Nyhan syndrome and PRPS over action.

Interesting regulation of purines

IMP formed can branch into either AMP synthesis or GMP synthesis. A conspicuous crosstalk is noted between the two paths. Understand this that energy is provided by GTP for AMP synthesis and ATP is used for GMP production. AMP and GMP both inhibit the expensive pathway. PRPP amidotransferase is slowed down by the presence of either of these while enzyme becomes fully inactive when both are present.

Chapter 19

Pyrimidine structure-2

Pyrimidine ring is an aromatic heterocycle of two nitrogen and four carbon atoms (Fig. 19.1). Apart from participating in constituting genetic material

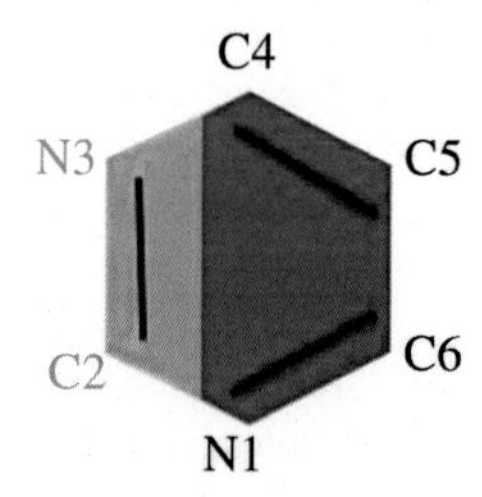

FIGURE 19.1 Pyrimidine ring.

DNA and RNA, some pyrimidines are found in nature, for example, vitamin B_1 and Riboflavin. Some drugs like anti-HIV drugs such as zidovudine and bacimethrin antibiotic are other examples.

Numbering of atoms is done in clockwise direction. Nitrogen atom is present at position 1 and 3. Source of carbon 2 and nitrogen 3 is carbamoyl phosphate, while rest of the ring is derived from aspartate. Only nitrogen 1 of pyrimidine ring forms glycosidic linkage with C-1' of ribose sugar (Fig. 19.2).

Sweet Biochemistry. DOI: https://doi.org/10.1016/B978-0-443-15348-8.00004-1

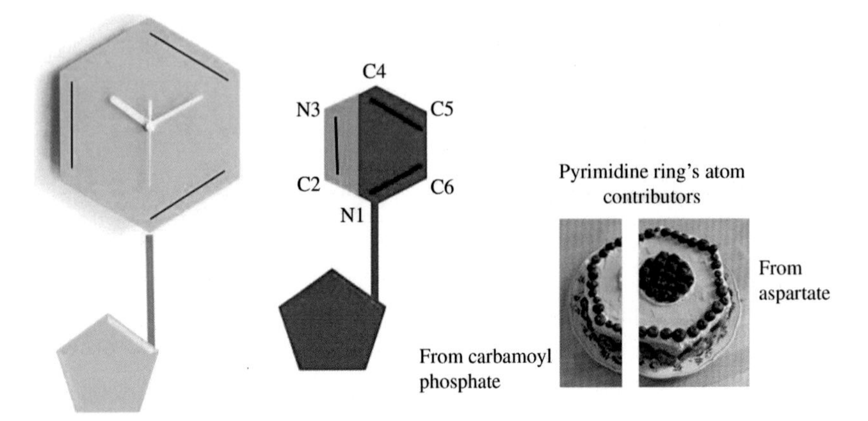

FIGURE 19.2 Pyrimidine ring can be compared with a clock. Numbering of pyrimidine is done clockwise and the sugar is attached to pyrimidine at N1 resembling pendulum hanging from clock. On right side, the contributors of Pyrimidine ring are depicted.

Uracil, cytosine and thymine are the principal pyrimidines which constitute uridine, cytidine and thymidine ribonucleosides and corresponding deoxynucleosides (Fig. 19.3). Uracil is chemically C4H4N2O2, thymine is

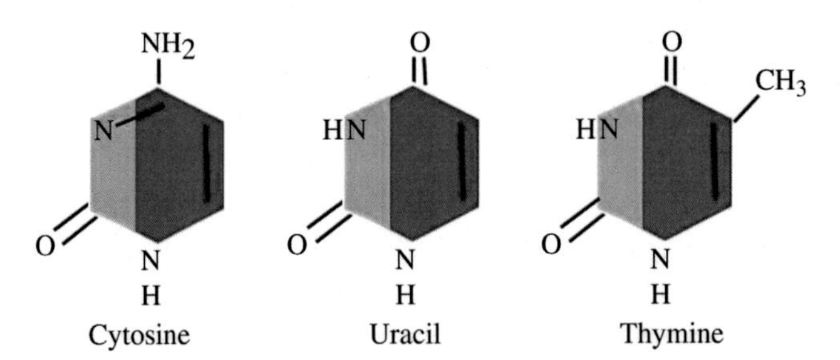

FIGURE 19.3 Important Pyrimidine bases.

C5H6N2O2 and cytosine is C4H5N3O. Cytosine and thymine are building blocks of DNA, while cytosine and uracil are found in RNA. Pyrimidine ring has a planar structure, and this helps in stacking interactions with purine bases. Pyrimidine pairs with complementary purine base by hydrogen bonding, for example, thymine with adenine and cytosine with guanine. Pyrimidines form not only building blocks of DNA and RNA, but also they serve important functions like polysaccharide and phospholipid synthesis, glucuronidation in detoxification and glycosylation of proteins and lipids. Structures of important pyrimidines are shown next.

Pyrimidines form not only building blocks of DNA and RNA, but also they serve important functions like polysaccharide and phospholipid synthesis, glucuronidation in detoxification and glycosylation of proteins and lipids.

Uracil

Uracil was found in 1885 while trying to synthesise uric acid derivatives; therefore, the German chemist Robert Behrend named it uracil. Five years later, Alberto Ascoli actually discovered this pyrimidine. An unique ability of uracil is that it can violate Watson–Crick base pairing rules. This enhances the RNA stability. It is a unsaturated compound with position of one double bond between C5 and C6. Uracil can absorb light. Tautomerism shift of amide-imidic acid form is displayed by uracil mainly near pH of 7. Amide form and imidic acid are called lactam structure and lactim structure, respectively. Out of these, first one is more common. Tautomerism is probably a compensatory mechanism for the absence of formal aromaticity (you know conjugated aromatic rings are stabilised by resonance), to increase nuclear stability. Uracil has nucleoside uridine and nucleotide uridine monophosphate (UMP), UDP and UTP. UDP-glucose is an important molecule.

Thymine

Thymine is 5-methyl uracil with similar unsaturated ring. Thymine name sounds like thymus gland, the source of isolation of this molecule was found in 1893 by Albert Neumann and Albrecht Kossel. Most important products of thymine are dTMP, dTDP and dTTP (sugar present is deoxyribose here). This is very interesting that uracil is present in RNA, while its methylated form thymine is seen in DNA. The explanation given for this is that cytosine can deaminate to form uracil which is repaired by DNA repair system. If uracil was already present in DNA, it would be very difficult for repair system to identify component uracil from deaminated cytosine.

Thymine is better resistant to photochemical mutation because it is more stable and less prone to damage. Hence, thymine replaced uracil in the prestigious DNA. Still thymine-thymine dimer formation is quite common on UV exposure, causing DNA kinks. This damage is repaired by photoreversal. 5-Fluorouracil targets thymine and is an effective cancer treatment.

Thymine binds adenine with two hydrogen bonds. In comparison to G-C base pair, the A-T base pair is weaker. This provides an advantage in placing A-T base pair where DNA strands should be opened easily during replication.

Cytosine

Discoverers of thymine found cytosine after 1 year, i.e., 1894. In the form of nucleotide cytidine triphosphosphate (CTP), this pyrimidine is a cofactor of many enzymes. Cytosine frequently undergoes spontaneous deamination, leading to uracil positioning in DNA. Thus, it is relatively an unstable base of DNA. This point mutation needs to be repaired. Cytosine methylation is catalysed by DNA methyltransferase, and 5-methylcytosine is formed. In epigenetics, this is of great value. Bisulfite sequencing is based on the difference of deamination rate of cytosine and 5-methylcytosine. When cytosine is present in codon at second position, third base is interchangeable. At third position, cytosine is read like uracil. G-C base pair is held by three hydrogen bonds. Greater bond strength, more resistance to desaturation and high boiling point are some of the implications of this.

Chapter 20

Pyrimidine de novo synthesis

Pyrimidine synthesis takes place in cytoplasm. Pyrimidine is synthesised as a free ring, and then a ribose-5-phosphate is added to yield direct nucleotides whereas in purine synthesis, ring is made by attaching atoms on ribose-5-phosphate. First three enzymes (blue-coloured) and fifth-sixth (green-coloured) are part of two multifunctional peptides to increase efficiency (Fig. 20.1).

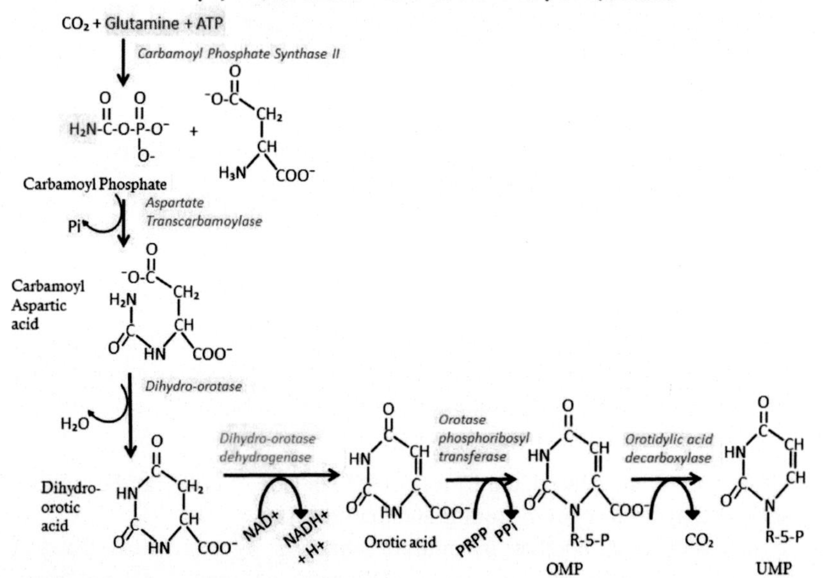

FIGURE 20.1 Pyrimidine de novo synthesis.

Important steps of pyrimidine synthesis are as follows:

1. *Carbamoyl Phosphate Synthase II step* – Carbamoyl Phosphate Synthetase II (CPSII) catalyses the reaction of bicarbonate and ammonia from glutamine in the cytoplasm to produce carbamoyl phosphate. This

Sweet Biochemistry. DOI: https://doi.org/10.1016/B978-0-443-15348-8.00011-9

enzyme is different from CPS I involved in urea synthesis. In the first reaction, glutamine reacts with CO_2 and two ATP to form carbamoyl phosphate. This step is similar to first reaction of urea cycle. In both reaction, carbamoyl phosphate is the product in the presence of enzyme CPS. But there are some differences you should remember here. First, the source of nitrogen in case of urea is ammonia (mostly coming from oxidative deamination of glutamate), while for pyrimidine synthesis is amide group of glutamine. Second, the CPS-1 is the mitochondrial isoform predominately expressed in liver (making it major organ for urea synthesis), while CPS-II isoform is active in cytosol of all cells. Third, the activators of both enzymes are different. For CPS-1, N-acetyl-glutamate is an indispensable allosteric activator and for CPS-II, ATP and PRPP are activators, while UTP inhibits the activity.

2. *Aspartate Transcarbamoylase step* – second main source of pyrimidine ring aspartate combines with carbamoyl phosphate in the presence of aspartate transcarbamoylase enzyme. This step is the committed step of pathway as this enzyme is allosterically regulated (allosteric inhibition by CTP).
3. *Dihydro-orotase step* – Covalent bonding between N3 and C4 closes the ring, yielding dihydroorotate. The enzyme participating is dihydro-orotase. These three enzymes are together called CAD, and it is a multi-functional protein.
4. *Dihydro-orotase dehydrogenase step:* Double bond between C5 and C6 is formed by enzyme dihydroorotate dehydrogenase utilising NAD + as coenzyme. Orotic acid is formed in this reaction.
5. *Orotase phosphoribosyl transferase* – Orotic acid is converted to orotidine monophosphate (OMP) by enzyme orotate phosphoribosyl transferase. Here, ribose-5-phosphate from PRPP is attached to N1 of orotic acid releasing pyrophosphate.
6. *Orotidylic acid decarboxylase* – Decarboxylation of OMP is catalysed by orotidylic acid decarboxylase. Carbon 7 of ring is removed as carbon dioxide yielding uridine monophosphate (UMP). Last two enzymes are also present as multifunctional protein.

UMP can be phosphorylated to form UDP and UTP. CTP can be synthesised by adding amino group from glutamine to UTP.

It is easier to remember pyrimidine synthesis by structures. Do you remember the sources of ring atoms: carbamoyl phosphate and aspartate. First prepare small molecule, that is, carbamoyl phosphate (Fig. 20.2).

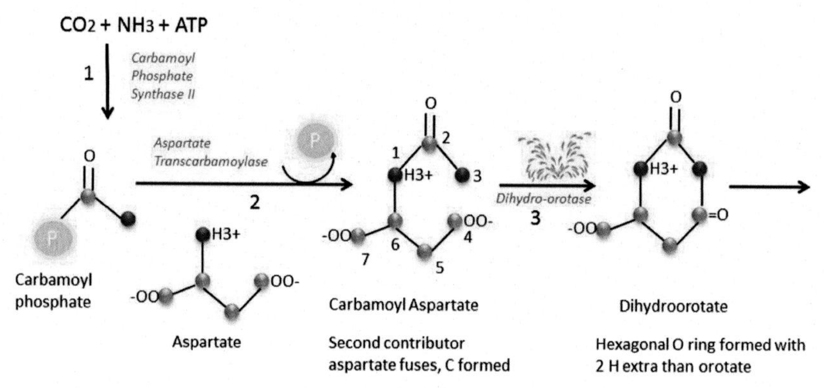

FIGURE 20.2 Simplified pyrimidine de novo synthesis-1.

As the name indicates, it generates from carbon dioxide (carb), ammonia (amo) and ATP (phosphate). Ammonia in this reaction comes from glutamine. Try to make a ring with the substrates carbamoyl phosphate and aspartate. They join to form a C-shaped open ring. Join it by dehydration. Polish it by dehydrogenation. Add a diamond on the ring and cut the extra carbon projections. Ring is ready (Fig. 20.3).

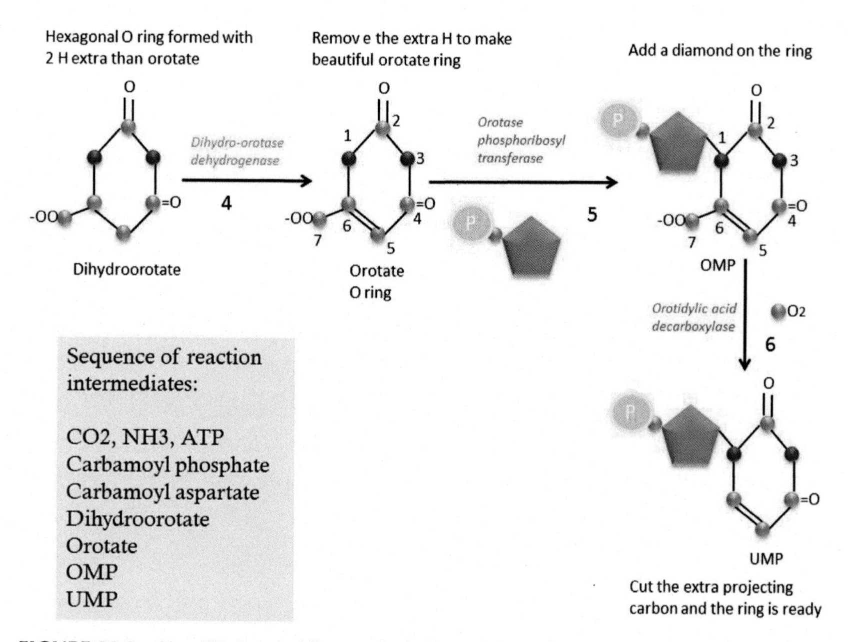

FIGURE 20.3 Simplified pyrimidine synthesis-2.

Regulation of pyrimidine de novo synthesis. Apart from allosteric activators and inhibitors of CPS II, phosphorylation modulates the pathway. MAP kinase can phosphorylate at two sites and decrease the UTP's inhibition and augment the activation by PRPP. Protein kinase A however by phosphorylation reduces the effect of both UTP and PRPP. ATCase enzyme is modified by cooperative attachment of carbamoyl-phosphate and aspartate. There are separate pools of carbamoyl phosphate in a cell's cytoplasm and mitochondria. When mitochondrial carbamoyl phosphate concentration raises due to urea cycle defects, it can cross mitochondria membrane and become a substrate of ATCase. In such conditions, PRPP becomes insufficient and UMP synthase becomes the limiting enzyme. This leads to orotate accumulation and excretion.

Chapter 21

DNA structure and DNA replication

Decoding DNA structure was one of the most priceless feats of science. The correlation of DNA sequence with the RNA and then the proteins continues to mesmerise the scientists all over the world. Let us delve in the intricacies of DNA molecule.

DNA structure is composed of two polydeoxynucleotide strands rotated about one another in double helix. In this helix, the N9 of purine or N1 of pyrimidine is linked by N-glycosidic bond to C-1' of deoxyglucose. On imagining, DNA as a ladder, its sides are formed by sugars connected by phosphodiester bonds and the nitrogenous bases linked by hydrogen bonds form the steps. Why ladder is mentioned for comparison because the strands are situated at a constant distance of 2 nm throughout the molecule. The strands are antiparallel that means the strands are placed in opposite directions. Hence, one is read in $5'-3'$ direction and vice versa.

Another feature of DNA is complementary nature of strands as adenine always base pairs with thymine and guanine associates with cytosine. This results in redundancy of information stored in DNA. The base pairs are positioned in a place perpendicular to the helix axis. Between consequent base pairs, 0.34 nm distance between adjacent pairs of bases and 10 bp exist in one turn of helix, with repeats every 3.4 nm. The studies using space-filling models described two sets of grooves in DNA: (1) major groove which is wide and deep, and (2) minor groove that is narrow and shallow. These grooves become important by providing binding site to regulatory proteins and other nucleic acids.

DNA has been selected as genetic material due to its great stability, and the underlying reason is the presence of deoxyribose. At 2'C of deoxyribose, oxygen is lacking that is present in ribose sugar. This unique change increases the resistance of DNA to hydrolysis.

These details were compiled by Watson–Crick model of DNA known as B-form. This right-handed helix is physiologically significant version present in solution and intracellularly.

The DNA stacking is stabilised by forces like hydrophobic interactions, Van der Waals forces and hydrogen bonds. Two hydrogen bonds unite A

Sweet Biochemistry. DOI: https://doi.org/10.1016/B978-0-443-15348-8.00035-1

with T and three combines G with C. So it is quite easy to understand that the segments predominantly having GC content will be difficult to open and have higher melting temperature, while AT segments should be placed at the regions for easy DNA opening.

Alternative DNA structures

DNA helix exists in different forms in different conditions. A summary of important forms of DNA known as A-DNA, B-DNA and Z-DNA is presented in Table 21.1 and diagrammatically represented in Fig. 21.1.

TABLE 21.1 Important variations of DNA.

	A-DNA	B-DNA	Z-DNA
Observed in	Form seen on removal of water from solution of DNA	Standard form seen in cell and solutions	Repeating CGCGCG hexanucleotide found in solution with zigzag configuration
Shape	Broadest	Intermediate	Narrowest
Rise/bp	2.3 Å	3.4 Å	3.8 Å
Helix diameter	25.5 Å	23.7 Å	18.4 Å
Screw sense	Right-handed	Right-handed	Left-handed
Glycosidic bond	Anti	Anti	Alternating anti and syn
Bp/turn of helix	11	10.4	12
Pitch/turn of helix	25.3 Å	35.4 Å	45.6 Å
Tilt of bp from perpendicular to helix axis (degrees)	19	1	9
Other features	The groups in A-form binds with less water molecules hence favoured in dehydration, minor groove almost disappears	Vital molecule for storage of genetic information	Displays DNA flexibility, Z-DNA involved in some viral pathogenesis like poxvirus
Biological significance	No	Yes	No

Note that A DNA is widest helix, formed by dehydration. Z DNA shows zigzag strands and is left handed, also see CGCGCG hexanucleotide. B DNA is biologically active form.

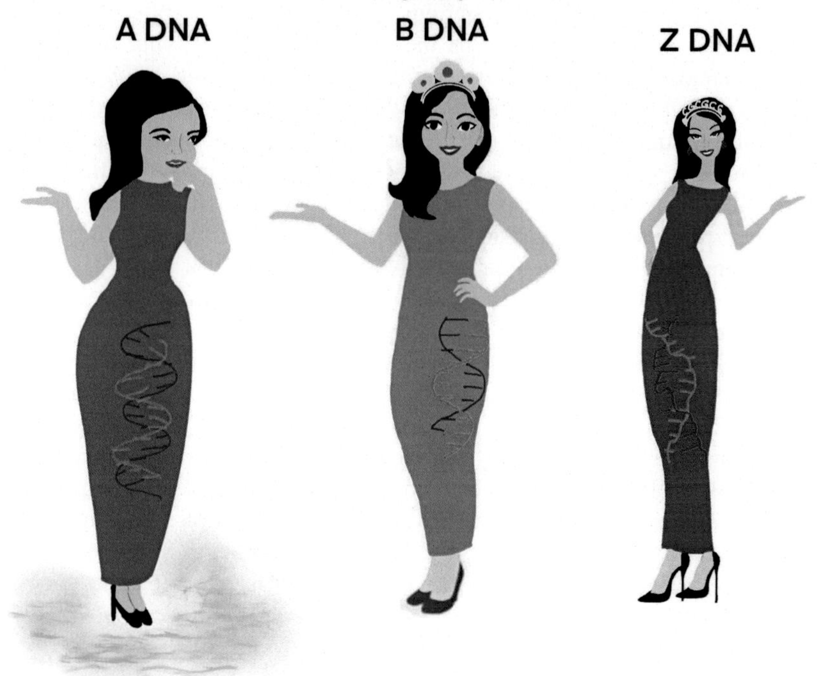

FIGURE 21.1 Types of DNA. The ladies are representing different forms of DNA. A DNA is formed by dehydration so water is seen near feet. Crown of cells is worn by B DNA.

Plasmid DNA: A circular small DNA molecule found in bacteria and mitochondria is known as plasmid. The replication of plasmid can be autonomous. Plasmid possess few genes to >100 genes out of which some are important like genes for antibiotic resistance and conjugation. It has been widely used as a tool in genetic engineering to carry and express foreign genes like those of insulin and growth hormone. Bacterias with modified plasmids have been grown into biological factories.

Triple helical nucleic acids: Felsenfeld and Rich were first to forward the concept of triple-helical nucleic acids in 1957. They described that a stable complex could be formed when polyuridylic acid and polyadenylic acid were present in ration of 2:1. The third strand was proposed to associate through the major groove of the dsDNA. Hoogsteen hydrogen bonds were responsible for the stability. Triple helix probability increases when a homopurine and homopyrimidine sequence were present in double strands.

When a mirror repeat sequence is present in dsDNA, one strand can interact as the third strand and allow formation of an intramolecular triplex or H-DNA. If triplex is pyrimidine-rich, it is known as H-DNA and if purine-rich, then *H-DNA.

DNA replication: The mysterious process of DNA replication which enables cell division and governs the life of almost all living forms is a unique interesting process. The action hero of this process is DNA polymerase discovered by Arthur Kornberg in 1958. Evolution has conserved the single polypeptide, 928 amino acids long with 109 kDa molecular weight structure. But the most conserved region of this enzyme is a tetrapeptide with two aspartate amino acid residue that probably binds magnesium.

DNA polymerase is envisioned as a right hand with thumb, finger and palm domains. Palm domain binds DNA, finger domain can attach the nucleoside triphosphate with template and thumb domain has a role in DNA positioning, translocation and processivity. Processivity means that the enzyme like DNA polymerase is adding nucleotides continuously on the template without detaching after each addition. The three activities of DNA polymerase are 3′–5′ exonuclease, 5′–3′ exonuclease and 5′–3′ polymerase. The actions may be easily guessed from the names. It is important to remember that DNA molecule is directional, and the two strands are antiparallel. Exonuclease action removes nucleotides and polymerase activity adds nucleotides. By 3′–5′ exonuclease function, proof reading, i.e., removal of erroneous nucleotide is done, while 5′–3′ exonuclease is targeted to remove RNA primer and DNA repair. Read the humanised dialogue of DNA polymerase in Fig. 21.2 and remember the functions.

FIGURE 21.2 DNA polymerase.

DNA polymerase catalyses phosphodiester linkage formation if pre-existing DNA template along with dATP, dGTP, dCTP, dTTP and Mg++ is available. A primer with 3′-OH group is needed to initiate the polymerisation. The 3′-OH terminal of growing chain performs a nucleophilic attack on phosphorus atom of deoxynucleoside triphosphate, resulting in the formation of phosphodiester bond and pyrophosphate release. The new DNA strand extends from 5′−3′ direction.

Brief summary of steps of DNA replication

A pre-replication complex is composed by initiation factors, origin recognition complex, Cdc6, Cdt1 and mini chromosome maintenance (Mcm) protein complex during G1 phase of cell cycle. Cell division cycle 45 (Cdc45) protein is fundamental for starting replication. Mcm also helps in loading of Cdc45 onto chromatin. Now this Cdc45 triggers loading of replication proteins including DNA polymerase alpha and epsilon and replication protein A onto chromatin. All this is to identify the replication sites.

Cdc6 then leaves forming the initiation complex. When cell enters, S phase of cell cycle, the complex becomes a replisome. The replication process starts in a chromosome at multiple origins during the S phase of cell cycle, with one origin being at approximate 30−300 kb of DNA. There are some replication factors like Claspin, And1 and replication factor C clamp loader that ensures that polymerase activity, DNA synthesis and unwinding of template goes hand in hand.

As the DNA helix is opened for replication, positive supercoils are formed for compensating decrease in numbers of twist. Supercoils are capable of stopping the replication. Therefore, topoisomerase removes the supercoils in front of progressing replication fork. Fig. 21.3 depicts important steps of eukaryotic DNA replication.

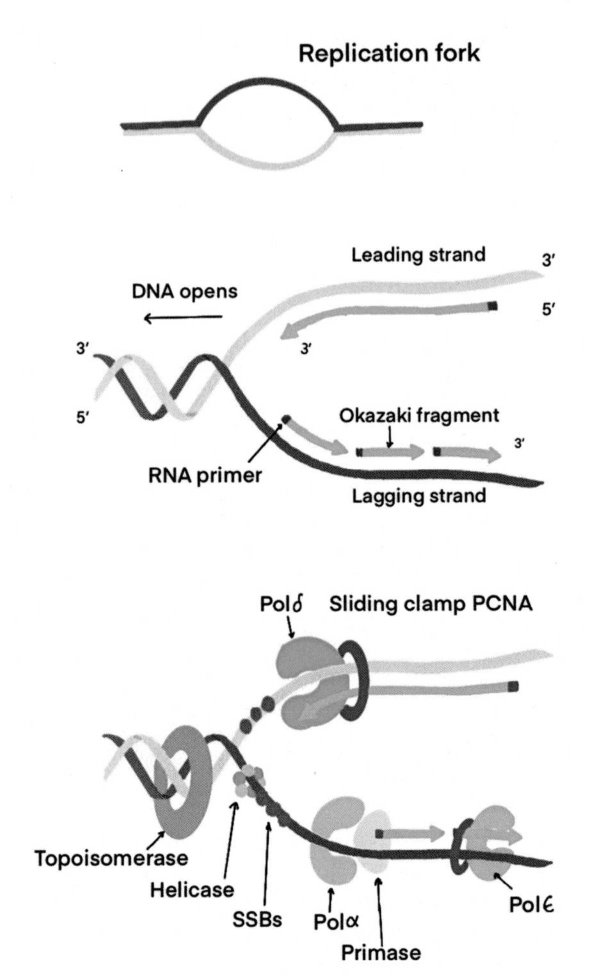

FIGURE 21.3 Glimpse of DNA replication in eukaryotes.

Out of two strands of DNA, leading strand grows in direction of replication fork movement while lagging strand synthesis takes place in direction opposite to fork. Former needs just one primer and is made continuously, but the later requires multiple RNA primers, and short fragments known as Okazaki fragments are synthesised. DNA pol α acts as a primase. Leading strand is made by DNA pol ε and lagging strand by DNA pol δ. Both strands are polymerised in $5'-3'$ direction and as two strands of DNA are antiparallel; hence, the synthesis of leading strand is little bit different from lagging strand. Lagging strand precursor Okazaki fragments are made by DNA pol at the $3'$ end of primer. The length of these can be 100–400 nucleotides long.

When the newly synthesised Okazaki segment reaches previous one, DNA pol δ displaces its 5′ end containing RNA primer and small DNA segment next to it. This RNA–DNA single flap is corrected by DNA pol δ and Fen1. Now only a DNA nick is left. Exonuclease action of Dna2 or Fen1 removes the primer. DNA ligase joins the DNA fresh segments together. Replication ends when two replication forks encounter each other.

Chapter 22

Differences in eukaryotic and prokaryotic replication

The process of replication have become complex along with the evolution from prokaryotes to eukaryotes. The requirements as well as regulators have increased multifold. To ensure the accuracy of replication and minimise mutation, species like humans have highly sophisticated and specialised enzymes for replication. In this chapter, we will distinguish between the replication of prokaryotic and eukaryotic genome. Have a look at Table 22.1 and Fig. 22.1 which give a quick glimpse at the differences.

TABLE 22.1 Differences between prokaryotic and eukaryotic replication.

	Stages	Prokaryotic replication	Eukaryotic replication
1. Initiation			
a.	DNA	Circular, double-stranded	Linear, double-stranded with ends
b.	DNA packaging	Prokaryotic DNA is wrapped on histone-like proteins resembling loops	Eukaryotic DNA is tightly wrapped on histones known as nucleosomes. Higher-order packaging like solenoid is also present
c.	Amount of DNA	Smaller amount	More than 50 times of prokaryotic DNA
d.	Occurrence	Prokaryotic replication is a continuous process	Eukaryotic replication takes place during S phase of the cell cycle
e.	Speed of process	Fast, around 2000 NTs added/second	Slow process, around 100 NTs added/second
f.	Site of replication	Cytoplasm	Cell nucleus
g.	Number of origin of replication	One per chromosome	Multiple per chromosome

(*Continued*)

Sweet Biochemistry. DOI: https://doi.org/10.1016/B978-0-443-15348-8.00033-8

TABLE 22.1 (Continued)

	Stages	Prokaryotic replication	Eukaryotic replication
h.	Size of origin of replication	100–200 nucleotides (in *E. coli*, ORI is 245 bp long and AT-rich)	Around 150 nucleotides. ORI of yeast has a specific sequence unlike humans where specific modifications to the nucleosomes recruit replication initiator proteins.
i.	Number of replication forks	Two	Several formed in numerous replication bubbles
j.	Number of replicon	One	One eukaryotic chromosome has >50,000 replicons
k.	Initiator molecules	Association of DnaA initiator protein to the DNA	Association of the MCM helicase with the DNA
l.	Unwinding of helix	Helicase that is a single hexamer	Helicase that is a double hexamer
2. Primer synthesis			
a.	Enzyme involved	RNA primase	DNA primase is present in a four-subunit complex, the DNA polymerase α/primase complex
b.	Primer length	5–10 NT-long primer	8–12 NT-long primer
c.	DNA polymerisation	DNA polymerase III	DNA polymerase δ
3. Leading strand synthesis			
a.	Removal of positive supercoils ahead of advancing replication forks	DNA topoisomerase II (DNA gyrase)	DNA topoisomerase II (DNA gyrase)
b.	Sliding clamp	β clamp in prokaryotes. Ring around DNA, ring around DNA is made up of two subunits of three domains each in β	Proliferating cell nuclear antigen (PCNA) in eukaryotes, the ring around DNA is composed of three subunits of two domains each in PCNA

(*Continued*)

TABLE 22.1 (Continued)

	Stages	Prokaryotic replication	Eukaryotic replication
c.	Molecules for stabilisation of unwound dsDNA	Single-stranded binding proteins	Replication protein A
4. Lagging strand synthesis			
a.	Okazaki fragments	Large, approximately 1000–2000 nucleotides fragment	Small, approximately 100–200 nucleotides in length
5. Primer removal			
a.	Enzyme for primer removal	Exonuclease activity of DNA polymerase I, and it also fills the gaps between Okazaki fragments in lagging strand replication	FEN1 (flap endonuclease 1) and RNase H
6. Ligation			
a.	Joining of Okazaki fragments	DNA ligase	DNA ligase
7. Termination			
a.	Replication at ends of chromosomes	Telomeres not present	Telomerase saves the ends of chromosomes
b.	Segregation of chromosomes	Segregation may start even before the completion of replication	Segregation starts when replication finishes in all chromosomes
c.	Product	Two circular chromosomes	Two linear sister chromatids

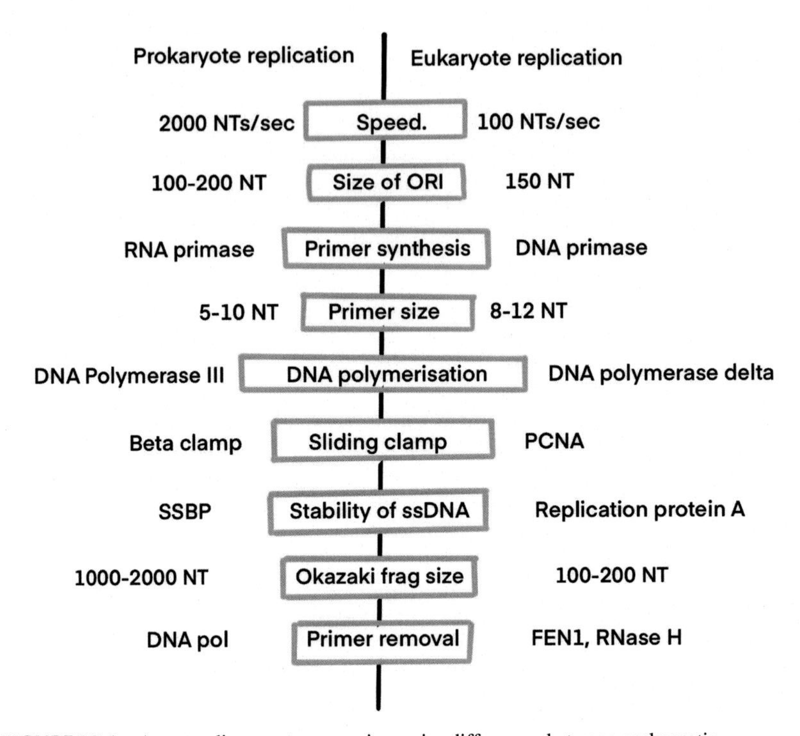

FIGURE 22.1 An easy diagram to memorise major differences between prokaryotic.

Chapter 23

Transcription

DNA is storing the information in nucleus. What is the benefit of this DNA? It is just lying there like a DNA caught in a test tube. The miraculous interplay of biochemical molecules emerges when this DNA is transcribed. Transcription literally means copying a piece of information. Copying is done in same language. So huge amount of data stored in DNA (deoxyribonuceic acid) is copied in parts in the form of RNA (ribonucleic acid) which is translated to proteins. Are you not curious to know about the key enzyme of this vital pathway? Name is derived purely from the function. RNA polymerisation – done by RNA polymerase (RNAP).

In prokaryotes, only one type of RNAP is present, having five subunits (alpha2, beta, beta', omega and sigma) in holoenzyme form which can polymerise and recognise the promoters, while core enzyme form which don't have sigma subunit can only polymerise. Sigma unit performs the sequence-specific recognition of promoter DNA. Eukaryotic RNAPs are more complex with 8–14 subunits. Beta and beta' subunits resemble the prokaryotic version.

How do RNA polymerases work? RNAP binds and melts 12–16 bp of the DNA strand at the promoter region forming what is called a 'Transcription bubble'. RNAP travels along the strand, catalysing the polymerisation in template-directed fashion. In terms of mechanism, RNA polymerases share brotherhood with DNA polymerases. In presence of Mg^{++} ions, 3′-OH of RNA primer performs a nucleophilic attack on the incoming rNTP. This adds one nucleotide at a time to the chain to the 3′ end of strand, so the chain elongates in the 5′–3′ direction. Eukaryotes have specialist RNAPs for processing different genes. RNAP I produces rRNA (28, 18 and 5.8S), RNAP II is involved in the synthesis of mRNA and microRNA, while RNAP III makes tRNAs and 5S rRNAs.

Steps of transcription

The first step of gene expression – transcription is divided into initiation, elongation of chain and termination.

Initiation: Bacterial transcription is possible in vitro when combination of purified RNAP and DNA with promoter is present. This is not the case in

Sweet Biochemistry. DOI: https://doi.org/10.1016/B978-0-443-15348-8.00012-0

eukaryotes due to additional requirement of transcription factors and sequences – TATA box and an initiator sequence. Fig. 23.1 presents the steps of initiation of transcription. Transcription factors have DNA-binding

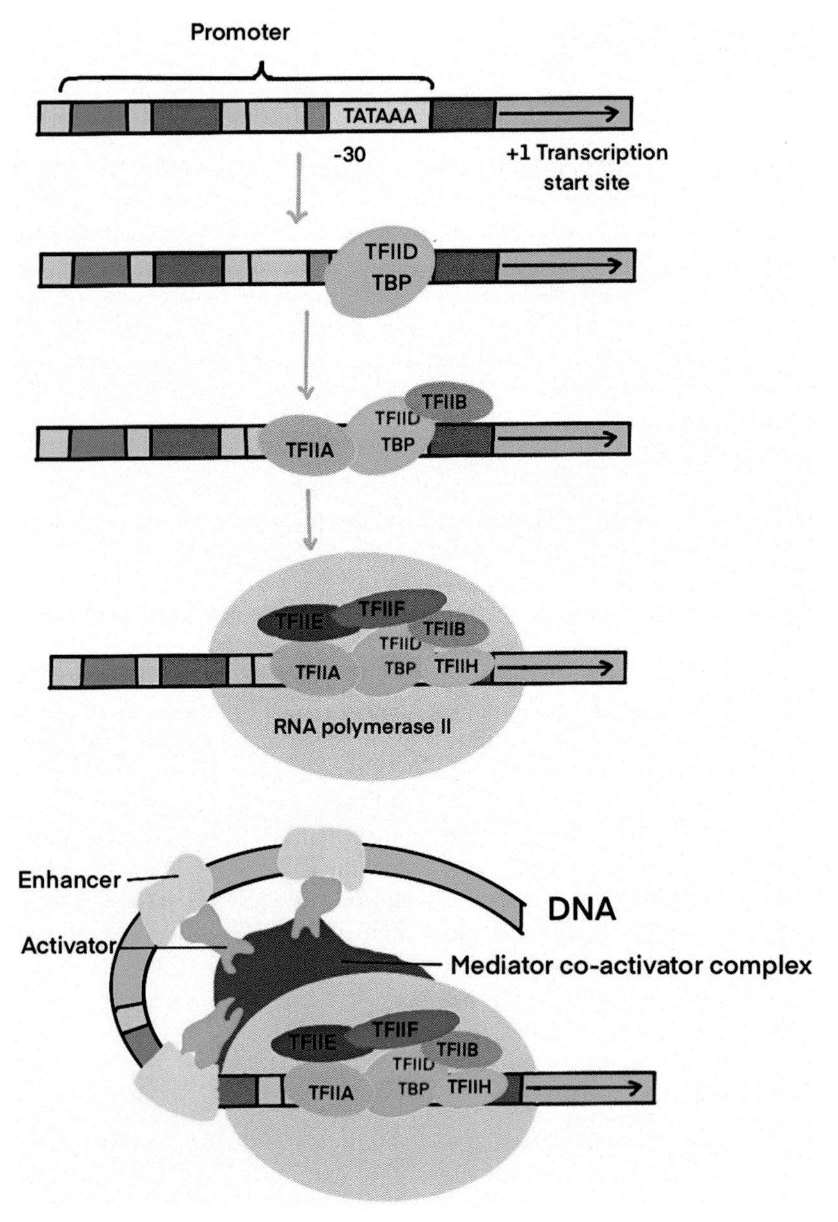

FIGURE 23.1 Initiation of transcription.

domains and can enhance or repress transcription process. Five general transcription factors needed are TFIID, TFIIB, TFIIF, TFIIE and TFIIH. TFIID is constituted by TATA-binding protein and TATA-binding protein-associated factors (TBPs) also called TAFs. TFIIB forms complex with TBP and RNAP. Another factor joins this – TFIIF. An interesting description of roles of various transcription factors is given in Fig. 23.2.

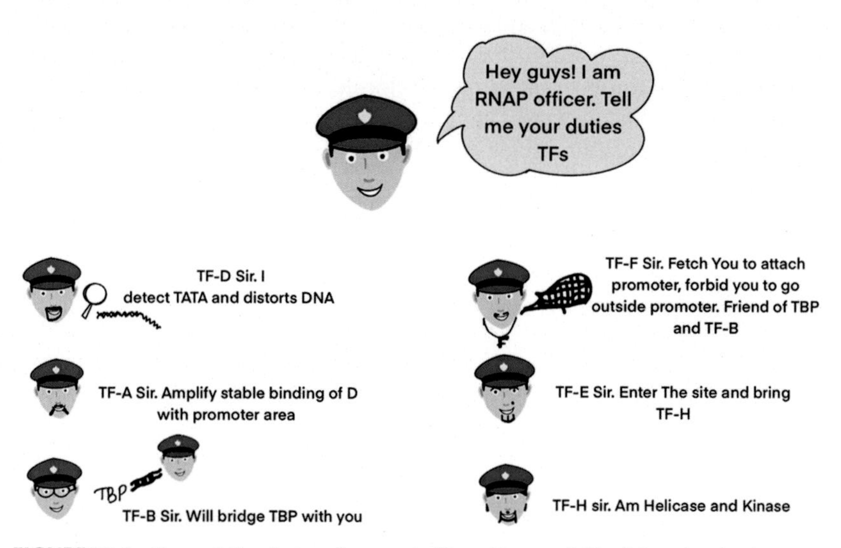

FIGURE 23.2 Transcription factors shown as traffic police squad. Read the roles slowly.

When RNAP attaches to promoter, TFIIE and TFIIH are needed. TFIIH has a helicase unit which helps in DNA unwinding and a kinase subunit that allows movement of RNAP on DNA template. Preinitiation complex of RNAP and transcription factors is formed at the promoter that is present upstream for every gene. Single-stranded DNA formed is ready to be copied. RNAP do not require any primer to start transcription. Rather it uses one strand of DNA known as template strand to guide ribonucleotide selection and polymerisation. The base pairing created by RNAP is A with U and G with C. In beginning, many abortive cycles result in hardly longer than 10-bp nucleotides. When the length of product increases, RNAP is able to cross this threshold, the enzyme is able to move ahead promoter.

Elongation: Beyond promoter, initiation factors are released and RNAP teams up with elongation factors. Out of which, positive transcription elongation factor (P-TEFb) is a prominent one. P-TEFb is a cyclin-dependent kinase that facilitates transition into productive elongation. Pre-mRNA nucleotides are quickly paired complementarily as per the template DNA sequence. Maintaining processivity, RNAP unzips the dsDNA and avails template strand. The equation of breaking one base pair in dsDNA and forming one base pair between DNA and RNA strand is followed. There are separate exits for nascent RNA chain and DNA strands. Behind the trailing end of transcription bubble, two strands of DNA again combine. Many protein factors are ready to process the upcoming mRNA by capping at the RNA exit channel.

Termination: The least understood last stage of transcription is termination. In prokaryotes, termination can be rho-dependent or rho-independent. Rho is a bacterial protein that attaches to the termination pause site. Eukaryotes transcription can be also factor-dependent or -independent. For RNAP II, factors like cleavage and polyadenylation specificity factor (CPSF) and cleavage stimulation factor (CSTF) can identify the poly-A signal in transcribed RNA and indicate termination. RNAP III can terminate the transcription without need of additional factors. Rather it pauses on recognising a short segment of polythymine on the non-template DNA strand. This stretch is present within 40-bp downstream from 3′ end of nascent RNA. The newly formed pre-mRNA is then released.

Chapter 24

Translation

Translation literally means conversion from one language to the other language. In molecular biology, translation represents the last step of gene expression, transforming the ribonucleotide polymer into a polypeptide chain of amino acids. Hence in this process, a chain of nucleic acids is converting into chain of amino acids. Isn't this is a wonderful translation? mRNA is translated from 5′ to 3′ direction. Prokaryotes have a comparatively simpler and faster translation process than eukaryotes. Differences have been highlighted in the end of this chapter. The model of translation described below is primarily prokaryotic. Translation is divided into three main stages:

1. Polypeptide chain initiation.
2. Chain elongation.
3. Chain termination.

To understand the stages better, let me briefly tell something about the ribosome because this organelle provides the framework for whole process, justifying the name protein factory given to it. Ribosome has three sites or slots where tRNA can attach in 50S subunit. The sites are called Acceptor (A) site that binds to aminoacyl-tRNA, Peptidyl (P) site which holds growing peptide chain and Exit (E) site. A tRNA passes from A site to P site and then uses E site to leave the ribosome.

Aminoacyl-tRNA synthetase/tRNA-ligase. The superspecificity of translation is totally dependent on aminoacyl-tRNA synthetase (aaRs) as it loads the tRNA molecule with an amino acid. Transesterification of a specific amino acid to its specific tRNA catalysed by this enzyme adds to the fidelity of expression of genetic information. You may ask how? A good question. Even if a tRNA has anticodon which matches with the codon on mRNA, if tRNA is carrying wrong amino acid, a mutated protein will be formed. Primary amino acids are those amino acids for which a tRNA is present. aaRS binds ATP and corresponding amino acid, forming aminoacyl-adenylate and inorganic pyrophosphate. The adenylate–aaRS complex interacts with the D arm of related tRNA, and amino acid forms a bond with hydroxyl groups at 3′ terminal of tRNA. This is how an amino acid is loaded on a tRNA molecule.

Sweet Biochemistry. DOI: https://doi.org/10.1016/B978-0-443-15348-8.00002-8

Initiation of translation: First step is assembly of 30-S ribosomal subunit, one mRNA, one charged tRNA-fmet, three initiation factors and GTP(Fig. 24.1).

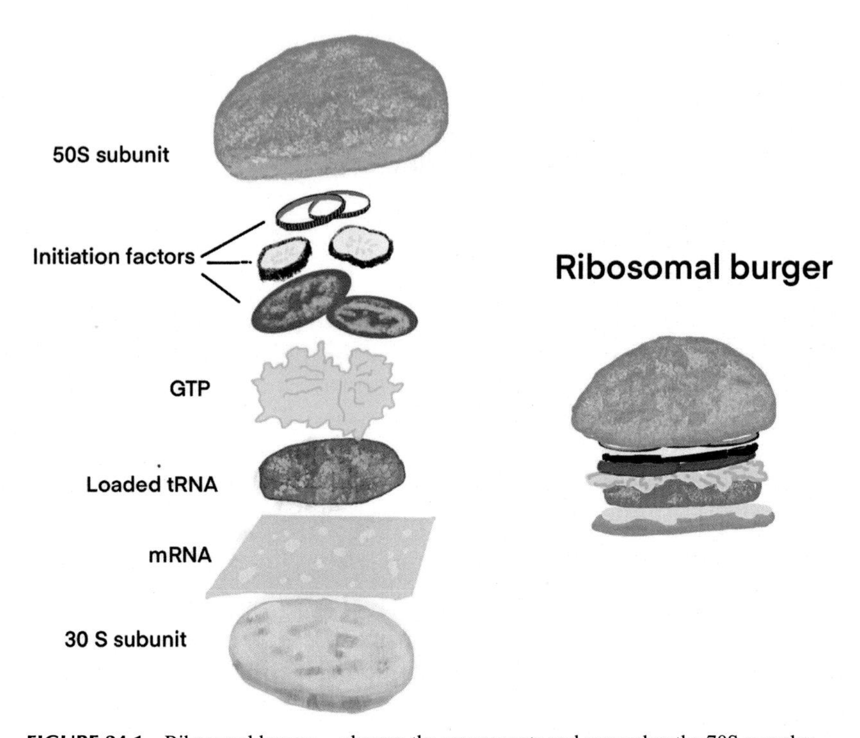

FIGURE 24.1 Ribosomal burger – observe the components and remember the 70S complex.

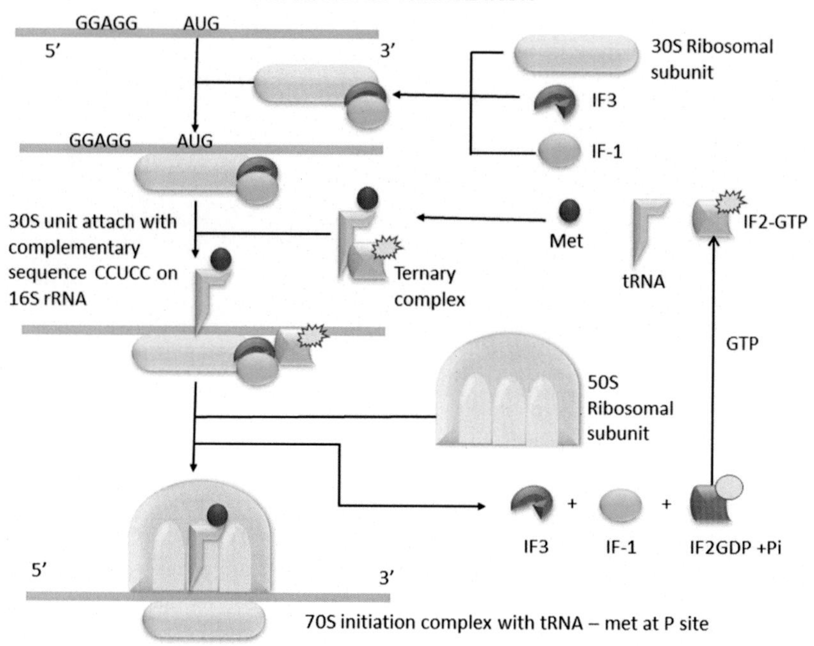

FIGURE 24.2 Steps of Translation-1.

This complex is known as 30-S preinitiation complex (Fig. 24.2). The tRNA associates with mRNA at start codon AUG. Then arrives the big subunit 50S. On its union which require energy from GTP hydrolysis, the complex is called 70-S initiation complex. The concept of ribosomal burger to remember 70-S initiation complex component is a really unique one. The 50-S subunit sits on 30S in a way that tRNA-fmet bound to mRNA is placed at P site. 'A' site is still empty and can be availed by a tRNA charged with the amino acid whose codon is present next to AUG. But this tRNA cannot enter like this. Steps of translation are shown in Figs. 24.2 and 24.3.

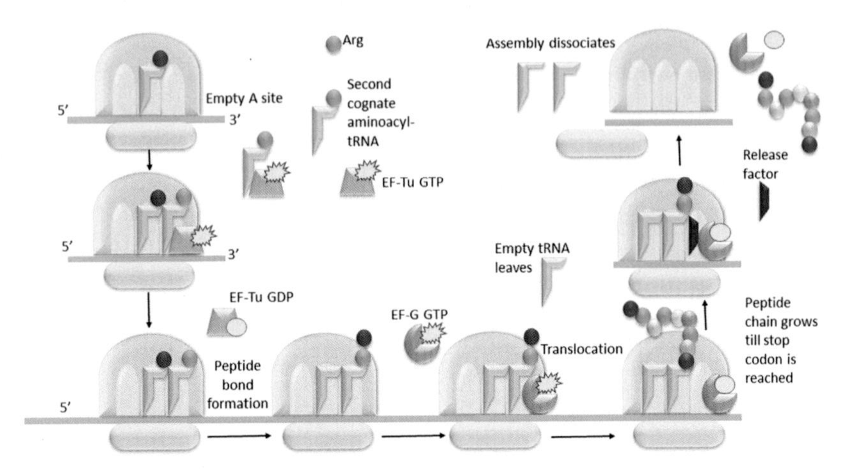

FIGURE 24.3 Steps of Translation-2.

Chain elongation. protein elongation factor EF-Tu binds GTP and an aminoacyl-tRNA (specific for the codon) and then this team enters A site. During the placement of aminoacyl-tRNA at A site, GTP hydrolysis takes place along with exit of EF-Tu-GDP. Two other elongation factors are EF-Ts and EF-G. EF-Ts acts as a guanine nucleotide exchange factor for EF-Tu, releasing GDP from EF-Tu. Now one tRNA is present at A site carrying an amino acid and another charged tRNA is at P site. Peptidyltransferase activity of rRNA creates the peptide bond between the two amino acids. Once a peptide bond is formed, methionine residue is separated from tRNA-fmet in P site by tRNA deacylase (Fig. 24.3).

Translocation of ribosome: Dipeptide linked tRNA is occupying A site, while at P site tRNA is bereft of amino acid. Here, three things happen – empty tRNA leaves P site, from A site the peptidyl-tRNA shifts to P site and ribosome moves by three bases on mRNA to position the next codon at A site. Energy needed for translocation is given by GTP in the presence of elongation factor EF-G to move ribosome. EF-G leads to big conformational changes and thus results in translocation of the tRNA and mRNA down the ribosome at the end of each elongation round. You can note that for making one peptide bond, four high-energy bonds are hydrolysed (Fig. 24.3).

Refilling of A site: A site is again vacant for a charged tRNA with correct anticodon for matching codon. At this site, new tRNA comes and similarly the chain elongates.

Chain termination. Translation continues until a stop codon is reached. For stop codons, no tRNA is present; therefore, the chain elongation stops. Instead, the release factors help the polypeptide chain to get released. There are three release factors RF1, RF2 and RF3. First two facilitates peptidyl-tRNA hydrolysis during closing of translation. RF3 is actually a GTPase which favours recycling of RF1 and RF2. Ribosomes split again into subunits 50S and 30S. An easy way to memorise translation is given (Fig. 24.4). Differences between eukaryotic and prokaryotic translation is vital and summarised in Table 24.1.

Learning Translation easily

So, you want to remember the process of translation. It seems complicated but let us try to simplify. Any process should have a starting, progression and finishing. These 3 steps for Translation are :

Initiation Elongation Termination

Factors participating in these steps will be named relatively. Like
Initiation- Initiation Factors- IFs
Elongation- Elongation Factors- EFs
Termination- Release factors- RFs

But do you know how many important factors are there in these 3 steps. 3 Factors for each of 3 steps

IF-1
IF-2
IF-3

EF-Tu
EF-Ts
EF-G

RF-1
RF-2
RF-3

These molecules work only in there step and leave before the next step starts. No confusion. These factors facilitate the process by providing energy or stabilise the molecules.

Now one more thing we often confuse is how to remember starting point for translation
In prokaryotes- Shine Dalgarno sequence AGGAGG upstream of first AUG

In Eukaryotes- 7-methylguanosine cap at 5′ or Kozak sequence is identified.

Initiation complex can be remembered as a set
30S + mRNA + 3 IFs + tRNA-fmet

You can now have a look at the proper steps . Enjoy

FIGURE 24.4 Learning translation.

TABLE 24.1 Differences between prokaryotic and eukaryotic translation.

S. no.	Feature	Prokaryotes	Eukaryotes
1	Relation of transcription and translation	Both processes are continuous and simultaneous	Both processes are separate. Transcription takes in nucleus, while translation is cytoplasmic
2	mRNA maturation	5′ end of mRNA is immediately available for translation	mRNA is processed and transported to cytoplasm
3	Active ribosome composition	70S type = 50S + 30S, 50S have 5S and 23S rRNA, 30S have 16S rRNA	80S type = 60S + 40S, 60S have 5S, 5.8S and 28S rRNA, 40S have 18S rRNA
4	Ribosomes location	Freely distributed in cytoplasm	Ribosome usually attached to endoplasmic reticulum
5	mRNA and gene information	Polycistronic (mRNA have information of more than one gene)	Monocistronic (mRNA has information of single gene)
6	Translation initiation mechanism	Cap-independent	Cap-independent and Cap-dependent
7	Start sites of translation	Many start sites and Shine–Dalgarno (SD) sequences present in one mRNA	One start site only, near to 5′ region of mRNA
8	Start site sequence	SD sequence exists 8 NT upstream of start codon	SD absent, Kozak sequence present few NT upstream of start site, cap of mRNA also helpful in identification
9	Initiation	30-S subunit identifies SD sequence during initiation	40-S subunit recognises 5′ cap of mRNA during initiation

(*Continued*)

TABLE 24.1 (Continued)

S. no.	Feature	Prokaryotes	Eukaryotes
10	First tRNA	Formyl-Met-tRNA	Met-tRNA
11	Required initiation factors	IF1, IF2, IF3	eIF1, eIF2, eIF3, eIF4, eIF5A, eIFB, eIF6
12	Elongation factors	EF-Tu, EF-Ts and EF-G	eEF1 and eEF2
13	Translation speed	~20 amino acids/second	~1 amino acid/second
14	Release factors for termination	RF1, RF2, RF3	eRF1
15	mRNA stability	Short lifespan, few seconds-min	Short to long lifespan, few seconds-years
16	Factor preventing the association of ribosomal subunits in the absence of initiation complex	IF3	eIF3

Chapter 25

Inhibitors of translation

Pathogenic bacteria are targeted by drugs like antibiotics. Everyone wonders what is the mechanism underlying this killer effect? One of the commonest mechanisms is being a translation inhibitor. In the process of translation, antibiotics can interfere with the major steps like ribosome assembly, or function of tRNAs and factors required in translation. It is quite crucial to memorise the inhibitors of antibiotics which is nicely summarised in Fig. 25.1.

Let us memorise the inhibitors of translation. You remember two units of Ribosome: 30 S and 50 S.

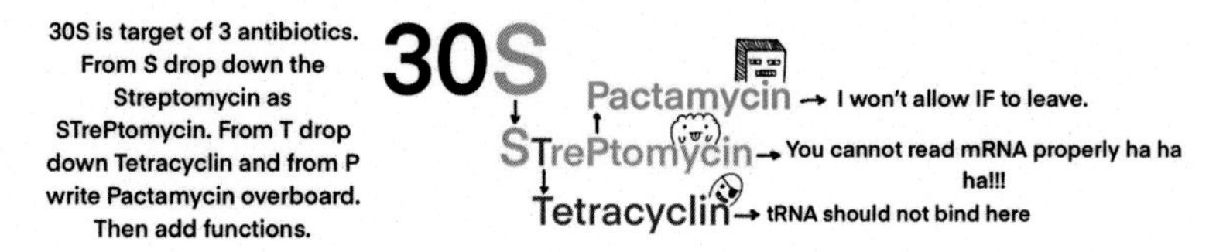

In the chapter you got some antibiotics with names starting from C and L. Do you recall an antibiotic with first letter E? Let's write 5 inhibitors of 50S and learn mnemonic CELL

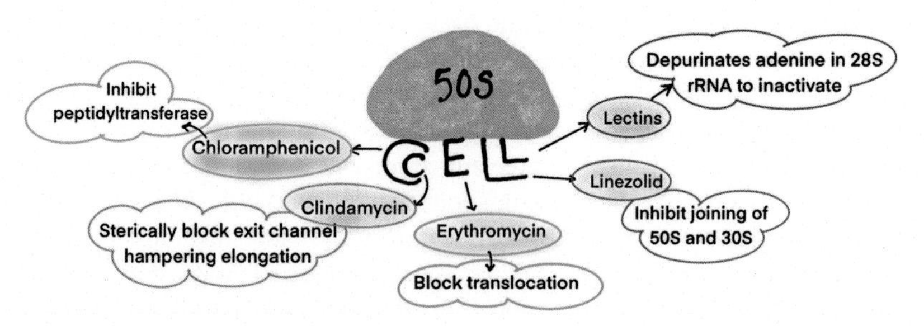

FIGURE 25.1 Inhibitors of translation.

Let's know about important translation inhibitors and the related pros and cons on health.

Sweet Biochemistry. DOI: https://doi.org/10.1016/B978-0-443-15348-8.00018-1

Antibiotics and the mechanism of action

Chloramphenicol. A small lipid-soluble molecule, capable of penetrating in all tissues, inhibits translation in bacteria and mitochondria at the step of peptidyltransferase activity. Chloramphenicol binds reversible to the 50-S subunit of 70-S ribosome particularly at A2451 and A2452 residues in 50-S unit. The interaction between amino acid containing end of aminoacyl-tRNA to the acceptor site on 50S is interfered. So the peptide bond formation cannot take place.

Cycloheximide is also known to inhibit peptidyltransferase activity of ribosome.

Erythromycin, Clindamycin: In comparison to chloramphenicol which hinders in substrate binding, macrolides are steric blockers for growing peptide chain. Erythromycin, for example, binds to 50-S ribosomal unit and interferes with translocation of peptide chain from 'A' site to 'P' site. Ribosome is not able to move along mRNA to show the next codon leading to premature termination of protein synthesis.

Streptomycin: Streptomycin is a broad-spectrum bactericidal aminoglycoside antibiotic which targets smaller subunit of ribosome, that is, 30S. Specifically, it attaches to the 16-S rRNA part of 30-S subunit and distorts the structure and interferes with attachment of formal-methionyl-tRNA. 50-S subunit association with mRNA is disturbed. These changes cause misreading of mRNA by alterations in the recognition site of codon–anticodon interaction. The frame shift of genetic information caused by streptomycin leads to halt in translation and cell death. Being a hydrophilic molecule, streptomycin is not able to cross cell membranes and need an electron transport system used during respiratory cycle of bacteria. This explains its potency in aerobic bacteria.

Tetracycline: First broad-spectrum bactericidal antibiotics are now less frequently used due to extensive bacterial resistance. Tetracyclines also bind to the 30-S ribosomal subunit and impede binding of aa-tRNA to the A site. It also prohibits the binding of release factors RF-1 and RF-2 during termination.

Puromycin: Puromycin is a molecule similar to 3′ end of aminoacylated tRNA with a modified adenosine associated with tyrosine. The difference between a aminoacyl-tRNA and puromycin is very interesting. In aa-tRNA, the bond between amino acid and the sugar ribose is labile ester linkage, while a stable peptide bond is present in puromycin. Like a normal aa-tRNA, puromycin enters the A site of ribosome and allows binding of its free amino group to the growing peptide chain at P site by peptidyltransferase. Now the problem emerges. The peptide bond present between two moieties of puromycin resists cleavage by an incoming aa-tRNA, so there is no chance of extension and so premature termination of translation leads to cell death.

Fusidic acid: Fusidic acid is a unique steroidal antibiotic produced by fungus Fusidium coccineum. It specifically inhibits elongation factor-G (EF-G).

EF-G possess GTPase activity. Fusidic acid binds to EF-G-GDP which is attached to ribosome and impedes translocation of nascent polypeptide along with hampering recycling of ribosomal subunits once the stop codon is reached.

Pactamycin: This drug is a strong inhibitor of protein synthesis in bacteria, archaea, and eukaryote kingdoms, so it is easy to think that it should act on some conserved region of 16-S RNA. Once pactamycin attaches to it, the release of initiation factors of translation is not allowed from smaller ribosomal 30S. Therefore, 70-S functionally active ribosomes cannot be formed. More specifically, pactamycin during initiation blocks factor and GTP-dependent binding of tRNA to P site of ribosome during initiation. This is an induced structural change in 30S and not any direct inhibition responsible for activity.

Another antibiotic Showdomycin prohibits the formation of eIF2-tRNA fmet-GTP complex.

Toxins that disturb protein synthesis

Diphtheria toxin: *Corynebacterium diphtheriae* produces a dangerous toxin named after it as diphtheria toxin that is a single polypeptide chain with 63,000 molecular weight and two intramolecular disulphide bonds. The toxin enters the cell and undergoes endocytosis mediated by receptors. The proteolysis of toxin generates two fragments – bioactive fragment A and larger fragment B. Role of fragment B is to help fragment A cross the cell membrane. Fragment A performs the catalysis of ADP ribosylation of EF-2 which is elongation factor. This modification turns Ef-2 inactive and hampers the protein synthesis by inhibiting the translocation. EF-2 is prone to this reaction due to the presence of an unusual amino acid diphthamide which is able to accept the ADP ribosyl group from NAD + .

Cholera toxin: Cholera toxin is another menacing molecule, a product of bacteria Vibrio cholerae. This toxin also mediates the ADP-ribosylation of guanidinium group of arginine residue present in guanine nucleotide-binding protein of adenylate cyclase. The result is the stimulation of adenylate cyclase. This enzyme catalyses the conversion of ATP into cyclic AMP. Elevated levels of cAMP promote the secretion of water and electrolytes from the intestinal epithelial cells. The extreme dehydration and electrolyte balance are therefore caused.

One thing you can appreciate that diphtheria toxin inhibits an enzyme, while cholera toxin stimulates to create the havoc.

Lectins

Lectins are proteins that bind carbohydrates specifically through carbohydrate-binding domain. These ubiquitous proteins are widely present in animals, plants, and microbial species. Examples of lectins are ricin, abrin and modeccin. Ricin is a heterodimer naturally found in castor beans and is highly dangerous for eukaryotes. It has two component polypeptides – Chain A and Chain B. Enzymatic Chain A leads to toxic effects, while Chain B attaches with galactose-containing proteins. Disulphide bonds connect these two components of ricin. B chain holds grip at cell surface by binding carbohydrates. A chain enters the cell by receptor-mediated endocytosis and depurinates one adenine at position 4324 of 28srRNA. This position is critical for binding translational elongation factors. This change makes ribosome inactive because the reaction is an N-ribohydrolase reaction.

Chapter 26

Operon

Everyone enjoys turning on air conditioners in the house during summers when suddenly the temperature rises. It is not wise to use AC in all seasons at all the times continuously. It is better to adjust it as per the requirement. Technology equips us to adjust with the environment changes. Another good habit is making kits, for example, first-aid kit. Kits ensure that all the needed material is present at one place so that no important component is left when we are doing that job. Bacteria also handle the variations in environment very smartly. In this chapter, let's see how?

Bacterial genes participating in a function are organised in clusters or coregulated genes. These closely placed genes are turned on or off together. This set of genes with common regulatory mechanism is called Operon and allows bacteria to find-tune its metabolism with the alterations outside. One of the best examples of Operon is Lac Operon present in *Escherichia coli*, including the enzymes for lactose metabolism.

The *lac* Operon

Francois Jacob and Jacques Monod noted that three genes coding protein enzymes for lactose metabolism form a cluster/Operon in *E. coli*. These genes are lac z, lac y and lac a, and these translate into beta-galactosidase, permease and beta-galactoside transacetylase, respectively.

Beta-galactosidase is an intracellular enzyme breaks lactose into glucose and galactose. Permease is basically a transmembrane pump which carries symport of lactose with proton. Beta-galactoside transacetylase is the enzyme which shifts acetyl group from acetyl-CoA to thiogalactoside.

Collectively, the expression allows entry of lactose and metabolises as an energy source. Along with sequentially placed genes downstream to a promoter, two important sequences known as operator and terminator are present. Promoter denotes the binding and starting site for transcriptional machinery. Operator sequence lies between promoter and structural genes, and if an inhibitor (lac i) binds here, whole Operon expression can be repressed. Inhibitor or repressor is encoded by regulatory gene i. Terminator sequence is the indicator of ending transcription by RNAP. Transcription of

Sweet Biochemistry. DOI: https://doi.org/10.1016/B978-0-443-15348-8.00034-X

Operon generates a polycistronic mRNA which produces multiple proteins participating in a pathway.

Lac repressor protein detects lactose indirectly by its isomer allolactose made by beta-galactosidase. It is synthesised constitutively. It's action stops when lactose is available. Another protein known as catabolite activator protein (CAP) is a glucose sensor. When glucose concentration is low, it stimulates transcription of Operon. CAP via cAMP estimates glucose levels.

Lactose is less preferred than glucose by *E. coli* as glucose metabolism needs less steps. But if lactose is the sole fuel available in environment, there is no option except to metabolise it by transcription of Operon.

When lactose is absent:

Repressor protein which is always expressed binds operator sequence and interferes with RNAP binding and thus Operon transcription. Lac genes expressed at very low levels in such condition.

When lactose is present:

Allolactose binds to repressor causing an allosteric change. The modified repressor can no longer bind to operator, and RNAP moves to transcribe the lac genes. Lac y gene code permeates in membrane which facilitates lactose inside the cell. Lac z product beta-galactosidase then cleaves lactose for metabolism. Now enjoy the cute poem on Operon in Fig. 26.1.

Operon Poem

Three genes in a line
One Operator, still that's fine

Before OP the Promoter lies
Ms. Poly have here professional ties

Up the lane reside the regulator
Always working and repressing the operator

But when lactose enters the zone
It stops the repressor from moving on

Operator gets free & Poly becomes happy
Enzymes are made as he fairly copy.

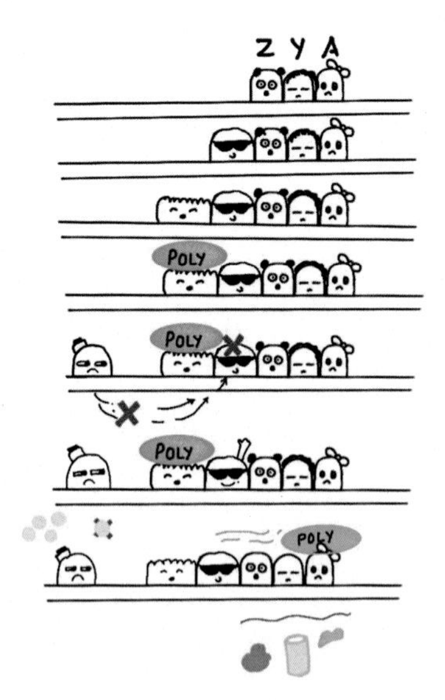

FIGURE 26.1 The Operon poem. Z, Y and A here are the genes beta-galactosidase, permease and beta-galactoside transacetylase. OP is Operator. Ms Poly is RNA Polymerase.

Chapter 27

Various types of RNA

Before discussing this multitalented unbranched single-stranded molecule, let us collect the clarity about the differences between DNA and RNA. So some important ones are shown in Table 27.1.

TABLE 27.1 Differences between DNA and RNA.

S. no.	Feature	DNA	RNA
1	Component sugar	Deoxyribose	Ribose
2	Component bases	Adenine, thymine, guanine and cytosine	Adenine, thymine, guanine and uracil
3	Base pairing	Adenine with thymine, guanine with cytosine	Adenine with uracil, guanine with cytosine
4	Number of strands	Double. Both strands intertwine around each other in form of a helix	Single mostly due to steric hindrance caused by the 2′-OH group of ribose. Helix not formed
5	Chargaff's rule (A = T, G = C)	Obeyed	Not obeyed
6	Modified bases	Absent	Present
7	Stability and half-life	Highly stable and hydrolysis by alkalis is difficult, longer half-life	Less stable and easily hydrolysed by alkalis, shorter half-life
8	Cellular content	Constant in all the cells except during cell division	Differ from cell to cell
9	Catalytic ability	Absent	Present

(*Continued*)

Sweet Biochemistry. DOI: https://doi.org/10.1016/B978-0-443-15348-8.00024-7

TABLE 27.1 (Continued)

S. no.	Feature	DNA	RNA
10	Present in	Nucleus and mitochondria (also chloroplast in plants)	Nucleus, mitochondria, nucleolus, ribosomes and cytosol
11	Length of molecule	Millions of nucleotides	Few thousands nucleotides
12	Role of genetic material performed in	All living organisms except some viruses	Few viruses like HIV

RNA can behave as a catalyst to a regulator of gene expression. In view of broad spectrum of RNA functions, RNA world was hypothesised. Intrahydrogen bonding leads to double-stranded RNA segments formation. Three major types of RNA are (1) messenger RNA or mRNA, constituting 5%–10% of total cellular RNA, (2) transfer RNA or tRNA that is 10%–20% of total, and (3) ribosomal RNA or rRNA is the rest, approximately 60%–80% of total. Eukaryotes also have one more family of RNA known as small-nuclear RNA (snRNA).

Messenger RNA

The secret message encrypted in DNA sequence is copied from one strand in parts in form of a messenger RNA. mRNA is most diverse set of RNA. It is complementary to the DNA sequence and therefore links the protein synthesis with the genetic information. This single-stranded RNA is the template according to which amino acids are joined to form a protein. Average length of mRNA in prokaryotes is 1.2 kb. RNA polymerase generates a transcript mRNA (or pre-mRNA) from DNA which after splicing converts to mature mRNA. 5′ end of mRNA possesses a 7-methylguanine triphosphate cap for protection against degradation. On finishing the primary transcript formation, a polyA tail is added to 3′ terminal which adds stability to mRNA and facilitates its transport. AUG is read as the start codon and translation stops at reaching stop codons like UAG, UGG and UAA. mRNA of prokaryotes is polycistronic which means that one mRNA molecule is representative of more than one gene, while in eukaryotes, the mRNA is monocistronic, i.e., having information of one gene only. Even the monocistronic mRNA is having introns (unnecessary segment) which are removed, and only exons are expressed. Eukaryotic mRNA is shown in Fig. 27.1. In prokaryotes, transcription and translation can proceed simultaneously unlike eukaryotes.

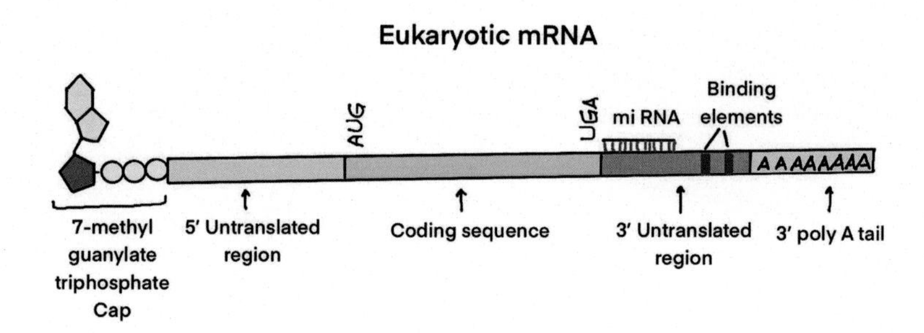

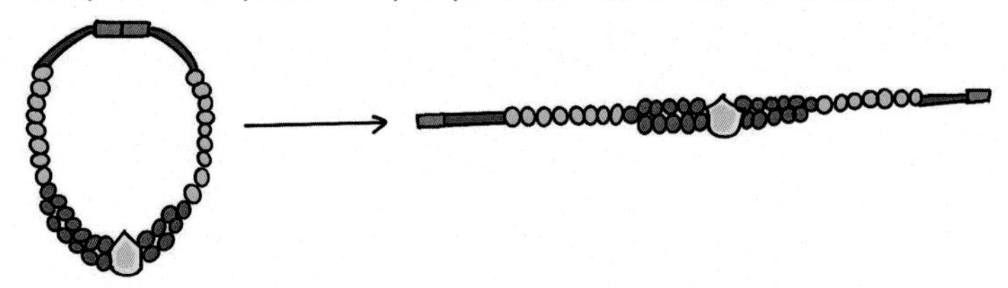

FIGURE 27.1 Eukaryotic mRNA.

Ribosomal RNA. rRNA is the core of ribosome in which protein synthesis takes place. rRNA contributes approximately half of mass of ribosome, while rest mass comes from proteins. But the rRNA is not just a structural addition, it performs the enzymatic role of peptidyltransferase and is therefore a ribozyme. Base pairing within rRNA sequences leads to stem-loop configurations which allows the formation of a three-dimensional structure. Two subunits have been identified – a large subunit and a smaller subunit. You will always see a capital S behind the numbers indicating subunit, for example, 50S. This "S" stands for size sedimentation coefficient. 50 and 30S refer to large and small ribosomal subunits in prokaryotes, respectively, while corresponding 60 and 40S are present in eukaryotes. Ribosome has three important sites known as E – Exit site, P – Peptidyl site and A – Acceptor site as shown in Fig. 27.2. Functions of these will be covered in Chapter 24.

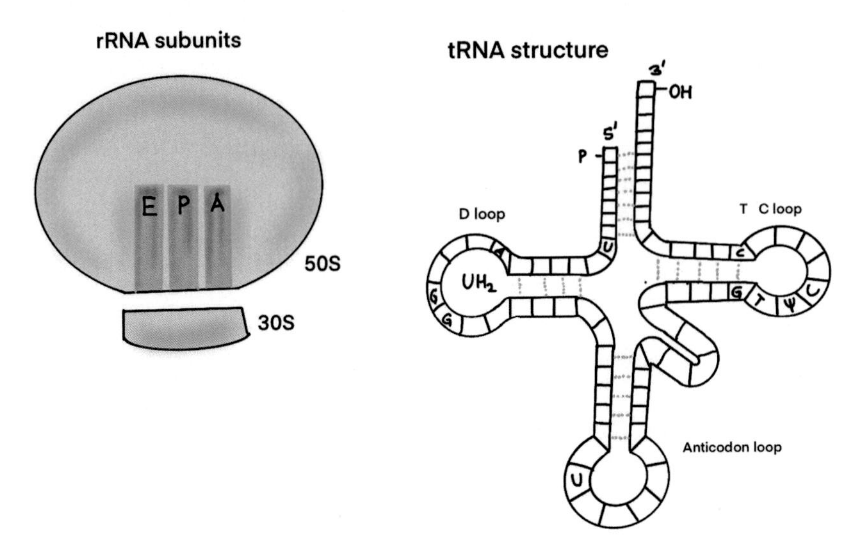

FIGURE 27.2 Structure of ribosomal subunits and tRNA.

Transfer RNA. It is abbreviated as tRNA, and Transfer RNA is an adaptor molecule which links sequence of nucleotides with sequence of amino acids. Isn't this amazing? With a size of 73–93 nucleotides, tRNA acts as an amino acid carrier; see Fig. 27.2. During the carrier service, tRNA is tagged with a superscript designating the amino acid. The presence of modified bases like dihydrouridine (DHU), ribosylthymine (rT), pseudouridine (Ψ) and inosine (I) enriches the molecules with special functions. Multiple double-stranded segments allows the formation of loops and stems. The secondary structure of tRNA resembles cloverleaf and in 3D, and it reminds of alphabet L. Base pairing at 5′-end (with phosphorylated base) yields stability to structure. tRNA has one acceptor stem, three single-stranded loops and a variable loop:

1. Acceptor stem – It accepts amino acid at CCA 3′ terminal group. Base pairing connects the 5′ and 3′ terminals. CCA (cytosine-cytosine-adenine) segment overhangs like a tail.
2. Anticodon loop – this loop preceded by 5-bp stem holds the anticodon. An amazing point to note here is that the anticodon is present in reverse order in 5′–3′ sequence because it reads 3′–5′.
3. DHU loop – this arm made from 14 to 21 bases varies in different tRNA and is named after the special base present in it – dihydrouridine.
4. TΨC loop – it carries modified uridine, pseudouridine.
5. Variable loop – a small loop with unidentified functions is present in some tRNA molecules.

Small-nuclear RNA

As the name highlights, this small ~150 nucleotides non-coding RNA is a family of RNA abundantly present in nucleus. You should wonder what it is doing there. snRNA is participating in intron splicing and pre-messenger (hnRNA) RNA processing. So like a scissor, it can cut RNA at required sites. snRNA associates with proteins to present as small-nuclear ribonucleoprotein complex which can act on bound unspliced RNA transcript. Human examples are U1 spliceosomal RNA, U2 spliceosomal RNA, and so on. U1, U2 and U6 are representing the uridine content. Additional functions of snRNA discovered are nuclear maturation of RNA transcripts, regulation of gene expression and maintaining the telomeres. The structure of snRNA is depicted in Fig. 27.3.

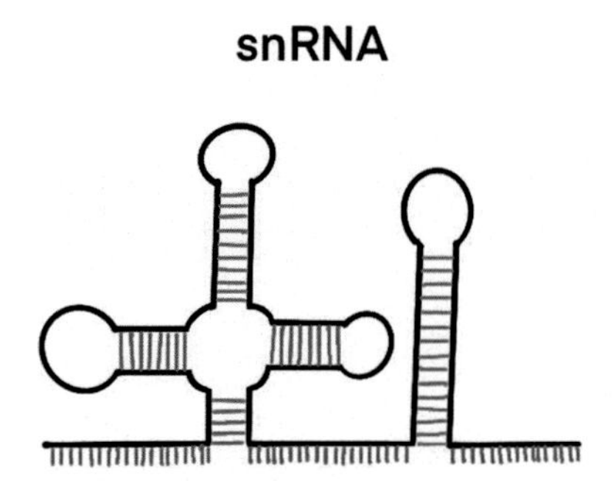

FIGURE 27.3 Small-nuclear RNA.

MicroRNA

It is quite clear that microRNA (miRNA) will be smaller than snRNA. Of course miRNA is smaller, i.e., only 19–20 nucleotides long. Moreover, it is single-stranded, evolutionary-conserved, non-coding RNA that plays important role in gene expression. miRNA suppresses translation by binding to the untranslated region on mRNA. So it has to be complementary to the target mRNA sequence. Gene expression can be unregulated also by binding of miRNA to promoters of gene. RNA polymerase II generates primary miRNA from processing of introns of host genes. Then, endonuclease Drosha and Dicer form the mature miRNA. miRNAs are used as biomarkers of diseases including cancer. The process can be seen in Fig. 27.4.

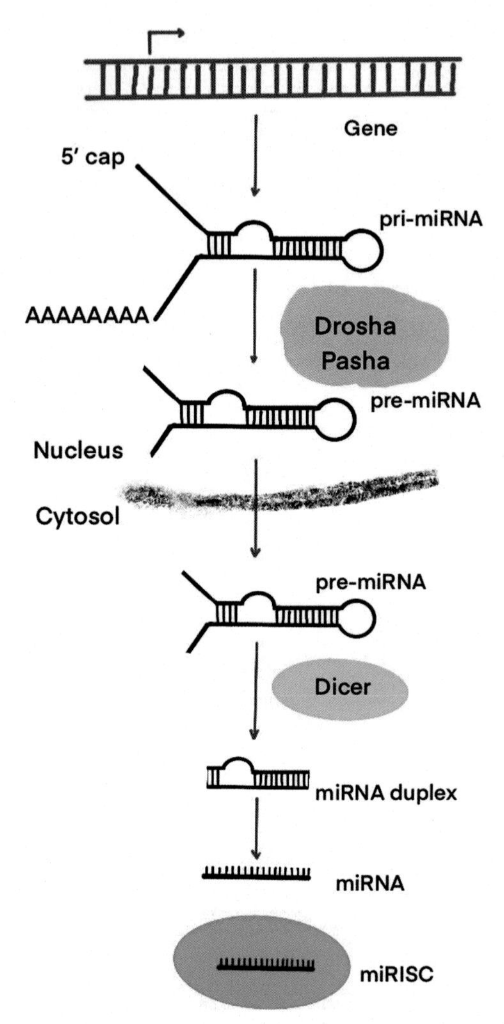

FIGURE 27.4 Micro RNA genesis. miRNA are transcribed by RNAP II & III. Primary miRNA (Pri-miRNA) is formed by transcription of miRNA gene. In nucleus, it cleaves into pre-miRNA (precursor miRNA). Drosha and Dicer are two ribonuclease III endonucleases. dsRNA is opened to yield ssRNA miRNA. RISC- RNA induced silencing complex is a multiprotein complex that incorporates 1 strand of miRNA or siRNA.

Small-interfering RNA

Small-interfering RNA (siRNA) is a class of RNA which is indistinguishable from mRNA biochemically and functionally. Both are same in size and have 5′-phosphate and 3′-hydroxyl ends. siRNA also joins RISC to silence a gene expression. miRNA can be differentiated from siRNA on the basis of origin. miRNA originates from double-stranded region of a RNA hairpin precursor;

on the other hand, siRNA is derived from long dsRNA. siRNA can be used to target a gene specifically.

Ribozyme

Ribozyme is a class of catalytic RNA molecules. Peptidyltransferase activity of rRNA mentioned above is a good example of ribozyme. Ribonuclease P (RNAse P) is a complex of enzymatic RNA and protein that can cleave a RNA if it is linked with an external guide (this is a short complementary oligonucleotide).

Chapter 28

Antibody

Antibody or immunoglobulin is a Y-shaped globular protein involved in defence against pathogens (bacteria and viruses) and unwanted foreign substances. Proteins are large polypeptides which folds in various shapes for performing different biological roles and the functional or structural unit of protein is known as a domain. So, a protein can have more than one domain. Similarly the antibody structure is divided into well-defined domains. The simplified diagrams of antibody immediately remind of alphabet Y. Fig. 28.1 will be helpful in this. This unique structure is constituted by four polypeptides: two smaller light chains and two bigger heavy chains.

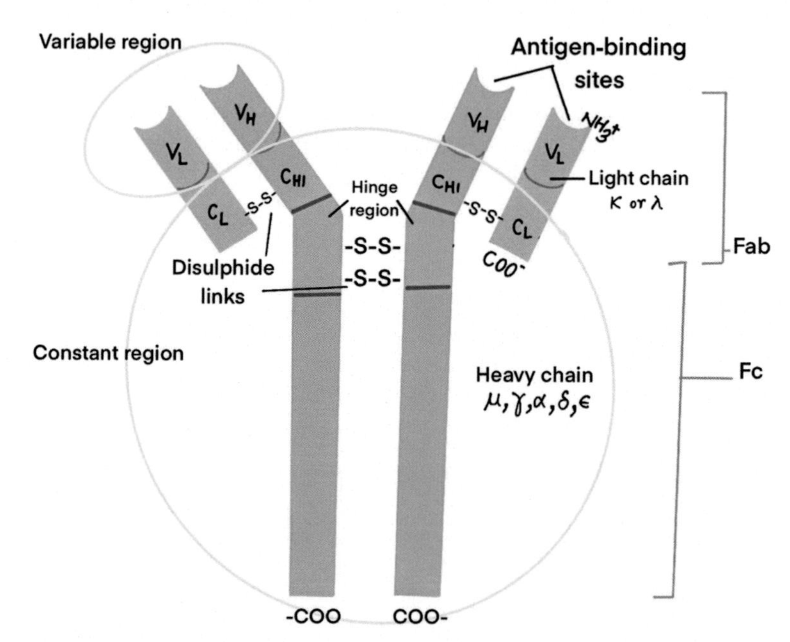

FIGURE 28.1 Antibody structure.

Each H-chain associates with a light chain, and the two heavy chains are bridged by disulphide linkage. Most important function of antibody, i.e., binding

Sweet Biochemistry. DOI: https://doi.org/10.1016/B978-0-443-15348-8.00014-4

of epitope by its paratope, takes place at N-terminal domain of heavy and light chains that constitute the antigen-binding site. This is upper V-like fragment (of Y-shaped antibody) which is also called Fab segment and lower single leg is the Fc portion. Amino acids variations are observed at this segment in different antibodies; hence, this is the variable site. More specifically, 6–10 amino acid residues are changed in the hypervariable part of variable region. By combining the hypervariable regions of a H-chain and a L-chain, Ag-binding site becomes specific to a molecule and is referred as idiotype determinant.

A very suitable analogy of antibody arms with tools used by ceramic artists is presented here. For doing unique actions on clay in pottery, only the tip of a tool is modified and rest the handle design is same for most of the tools. Similarly, nature did not design whole antibody structure to be variable. Greater proportion of antibody structure is constant and performs crosstalk with other immune cells. One domain in L-chain, the C1, is constant and can be of two types – kappa κ (60%) or lambda λ (40%). There is no functional difference between the two. In one Ab, either kappa or lambda is found. In heavy chain, 3–4 constant domains (CHs) are observed. Subtypes of CH are named after Greek letters mu μ, gamma γ, alpha α, delta δ and epsilon ε. These heavy chains decide the isotype or class and biological functions of antibody. So, when we say IgM antibody, it means that mu μ constant heavy chain is present. On binding of Ag at the paratope, a conformational change is induced in the heavy chain for enabling its biological functions with respect to its class. The characteristics of these five antibodies are provided in Table 28.1. Visual mnemonics to help in memorising important details of antibodies are given in Figs. 28.2–28.4.

TABLE 28.1 Antibody types present in body.

Important features of immunoglobulins classes					
	IgM pentamer	IgG monomer	IgA secretory dimer	IgE monomer	IgD monomer
Heavy chain present	μ	γ	α	ε	δ
Antigen-binding site present	10	2	4	2	2
Molecular weight	9,00,000	1,50,000	3,85,000	2,00,000	1,80,000
% of total Ab concentration	6%	80% (highest concentration)	13%	0.002%	1%
Placenta crossing	No (due to large size)	Yes (provides passive immunity to foetus)	No	No	No
Complement binging	Yes, excellent due to five complement binding sites	Yes	No	No	No
Fc role		Phagocytes		Mast cells, basophils	
Functions	First Ab secreted in blood and lymph for primary immune response, good fixation of complement, monomer unit acts as a receptor on B cell	Predominant action in secondary response, activates classical pathway of complement system, neutralise toxin, role in Ab-dependent cell-mediated cytotoxicity	Protects mucosal surfaces present in mucus, tears, saliva and colostrum	Ab responsible for allergy and fights against parasites especially helminths and protozoa	B-cell receptor

Follow the steps to remember important features of IgM & IgG

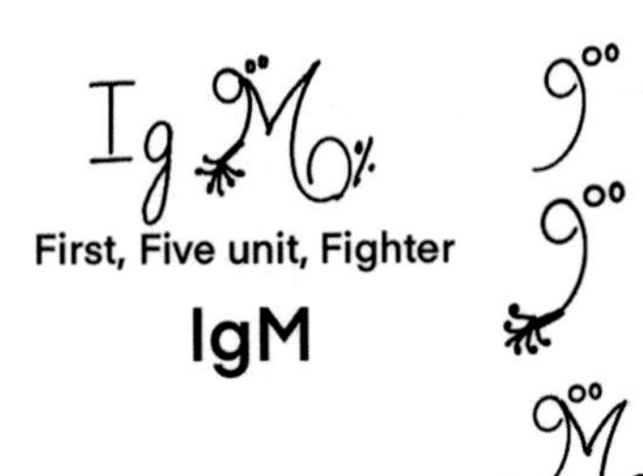

Draw a 9 and two small zeros, this shows mol.wt of IgM I.e. 900 thousand.

Draw a flowery entry in tail of 9 representing structure of C1 complement that is activated by IgM.

Draw the curvy second arm of M in form of a 6 and write small % with it. This is the % of IgM of total Ab in plasma.

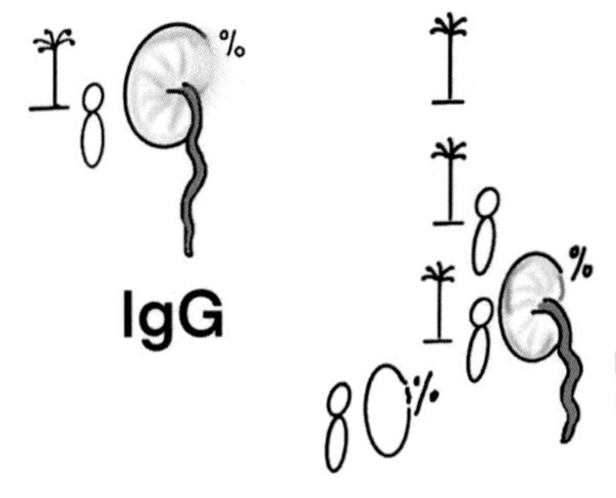

Draw I with a flowery top that reminds of complement C1 activation

Draw g in elliptical form resembling 8. This is useful

Draw G like the diagram. Here G alphabet resembles placenta to show that IgG crosses placenta. Also add % near top. Now see zero in circle of G. Read 8 from middle letter g. 80% of total Ab is IgG.

FIGURE 28.2 Visual mnemonic to learn IgG and IgM.

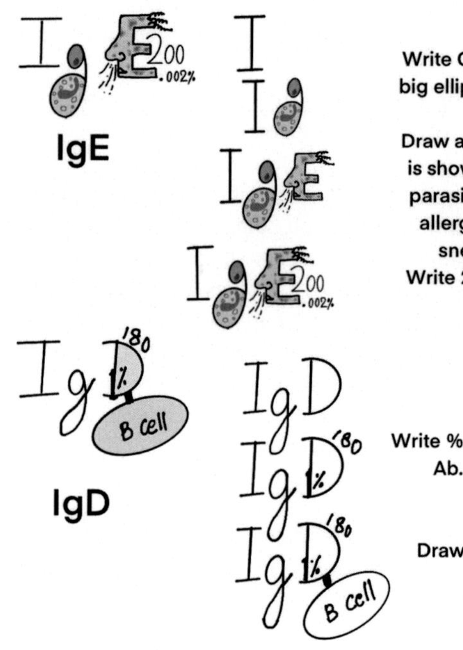

Write I

Write G and make its small circle mast cell & big ellipse the basophil. Fc of IgE binds these 2 cells.

Draw an interesting E as shown. Top line of E is shown with worms leaving bcz IgE is anti parasitic. Colour of E is with red patches of allergy caused by IgE and nose is shown sneezing because of allergic rhinitis.

Write 200 I.e. molecular weight and 0.002% I.e. % from total Ab.

Write IgD

Write % with line of D. 1% is fraction from total Ab. 180 is showing molecular weight.

Draw a B cell with whom IgD is attached because IgD is a receptor.

FIGURE 28.3 Visual mnemonic to remember IgE and IgD.

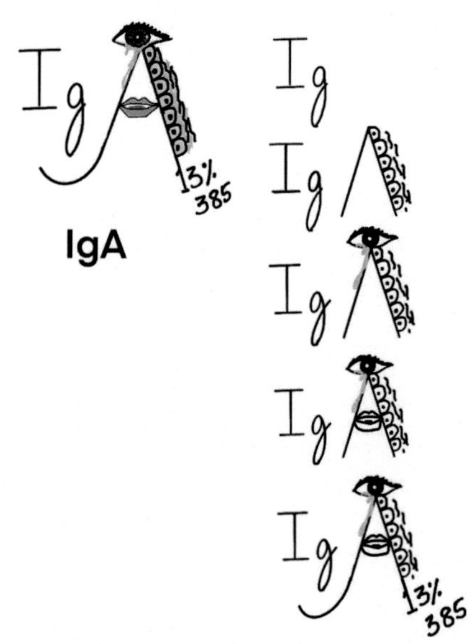

Write Ig

Draw sides of letter A & on one side draw some cells and add mucous to show mucosal secretion of IgA

On top of A , draw a eye with tears flowing to reflect presence of IgA in tears

Between two sides of A, draw a mouth as IgA is found in saliva also

Last write 13 % on one leg of A that is % of total Ab. 385 is molecular weight in thousand. Make other leg of A curved so that it looks like J. J chain is important component for secretion.

FIGURE 28.4 Visual mnemonic to learn IgA.

Antibody is capable of binding antigen. The extreme specific complementary binding between part of antibody with a part of antigen resembles an enzyme binding with its substrate. From previous chapter on antigen, we know that there are multiple epitopes on an antigen. These epitopes are bound by complementary paratope of antibody. Hence, one antigen can be attacked by numerous antibodies.

Fc tail with another name fragment crystallisable region is recognised by cell surface receptors called Fc receptors present on hematopoietic and effector cells and few complement system proteins. This interaction facilitates phagocytosis, endocytosis of Ab-tagged particles, releases of inflammatory molecules and antibody-dependent cellular cytotoxicity (ADCC). Fc region of immunoglobulin IgG, IgA and IgD is made of two same protein domains belonging to second and third CHs of Ab heavy chain. Longer Fc segment in IgM and IgE is formed by three heavy-chain CHs. A special conserved glycosylation at Asn 297 is noted in IgG-Fc that is involved in crucial interactions with effector proteins.

Chapter 29

Class switching

Antibodies present in our body are categorised in five classes, namely IgM, IgG, IgA, IgE and IgD. And switch, everyone is familiar with so many types of switches. Switches are used to change between two or more states. Therefore class switching should represent a change in class of antibody synthesis, for example, from IgM to IgG production. Other names for this amazing biological mechanism include isotype switching, isotypic commutation and class-switch recombination (CSR). Interesting fact here is that switching of only constant part of heavy chain takes place, while variable segment remains same. A curious mind may be wondering why this is so? Variable segment is involved in identifying antigen, and we want to target same antigen again. But constant region determines class of Ab which modulates effector actions as well as stability of molecule. Constant domains of heavy chain (CH domains) are bound by various cell surface receptors and complement system. Hence, class switching allows desired change in function.

Now **see the options for switching**: IgM, IgG1, IgG2, IgG3, IgG4, IgA1, IgA2, IgE and IgD. Genes for these total nine antibodies are available. What should be the sequence of antibody expression? The sequence of heavy chain exons decided by nature is IgM, IgD, IgG3, IgG1, IgA1, IgG2, IgG4, IgE and IgA2. Have you noticed how a chef places the storage containers in his/her kitchen? The most frequently used item will be placed nearer to the hands, and the rarely used items are placed on upper shelves behind the containers. From antibody functions, the sequence seems to follow similar logic. IgM being the first antibody for primary immune response is synthesised first and the IgD is acting as B cell receptor. B cells are mediators of Humoral response. IgG plays major role in secondary response. IgG1 and IgG3 activate complement very effectively and tackles most of the protein antigens. IgG2 and IgG4 are able to cater carbohydrate antigen mainly, and the opsonisation done by these two is also poor. After IgG, comes another abundantly present antibody of serum - IgA which has been recently shown to regulate immune responses also. IgA is composed of IgA1 and IgA2, former constituting almost 90% of total. IgA1 is anti-inflammatory while IgA2 is pro-inflammatory. It appears that entry of IgA1 is an attempt to limit the spread of inflammation during immune response while when an attackers persists long, IgA2 comes later to induce

Sweet Biochemistry. DOI: https://doi.org/10.1016/B978-0-443-15348-8.00003-X

cytokine production. IgE is placed at back positions as the functions is more specific in location or antigen attacked.

When is class switching required?

Immature B cells who have never encountered an antigen express only IgM immunoglobulin in monomeric cell surface–bound form. During maturation, IgD expression also starts and the cell is ready to identify antigen. On interaction with antigen, the cell division and differentiation of B cell into plasma cell take place. These plasma cells secrete antibodies which in beginning are IgM.

Here comes the role of mediators of class switching – the cytokines that are identified by CD40 and cytokine receptors of activated B cell. Cytokines are released by T helper cells, and now induction of plasma cells to secrete other suitable classes of antibodies like IgG, IgA and IgE takes place.

Mechanism of class switching

Now understand the mechanism of class switching. In an Ab, there are two chains – light chain and heavy chain. Heavy chain determines the class of Ab; therefore, here the concern is heavy chain of Ab only. Antibodies are globulin proteins, and proteins are coded by genes. Hence, one should look at the gene arrangement for heavy chains constant region. As mentioned before, variable region won't be changed.

The name CSR may help in understanding the process. Recombination is actually re-combination. But how this can be done at DNA level? For recombination, DNA has to be cut and then joined, then it will be called recombination. Now have a look at the chromosome 14 which hosts the genes for variable region and constant region of H-chain. Between variable segment gene (V) and constant portions lie two exons named D and J. For variable polypeptide, 1 V combines with 1 D and 1 J gene. Genes coding constant component of Ab lies in the order of Ab expression mentioned above.

Do you think it is easy to cut adjoining genes precisely without any damage to information. Well I think it is better to add some introns between various genes. So the intrachromosomal deletional recombination takes place at or near 'Switch or S' regions. These Switch regions are G-rich tandem repeat rich, 20–80 bp DNA segments present upstream of CH region genes except C-delta. The CSR is end-joining type of recombination. In simple words, first an error is introduced in DNA at cutting site and an acceptor site which on repairing causes double-stranded break. Then DNA ligation is done to form desired functional genetic combination. You would appreciate the wonderful mechanism as shown in Fig. 29.1.

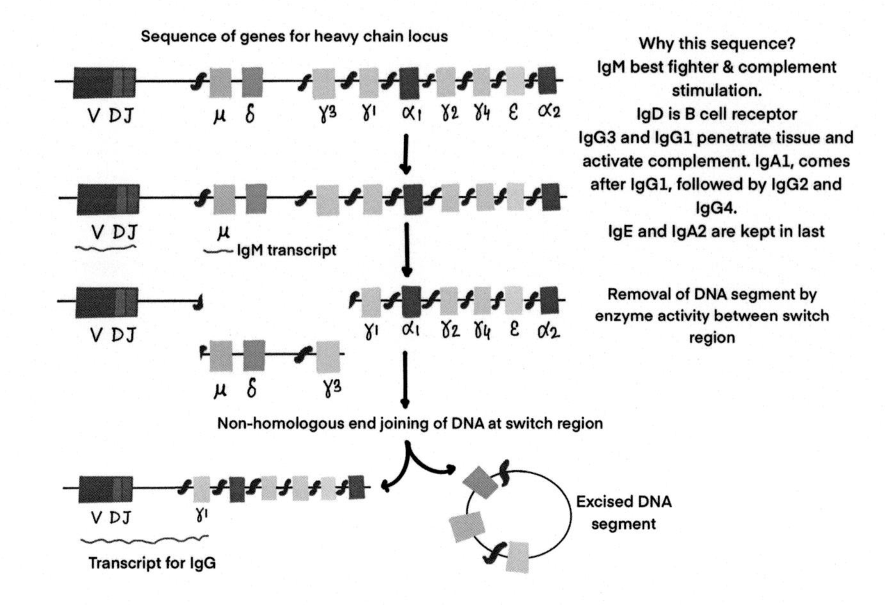

FIGURE 29.1 Class switching mechanism.

Now let me elaborate the process. During CSR, cytidine deaminase is stimulated that deaminates dC residues in switch region forming uracil. Uracil is abnormal at this place and is counted as a damage or mutation. Mismatch DNA repair and base excision repair proteins cause DNA breaks in both strands at two places. Take for example S-mu and S-gamma2b region. Double-stranded breaks are joined by nonhomologous end-joining. Now VDJ region unites with CGamma 2b gene, and extra segments are looped out. And the Ab formed is IgG2.

Chapter 30

Antigen

Antigen is a molecule recognised by components of immunity like antibodies or lymphocytes resulting in immune response generation against it. Another term immunogen needs to be differentiated clearly here. Antigen binds to components of immune response, and immunogen can induce immune response. So every immunogen is an antigen, but every antigen may not be immunogen. Molecular structure of antigen can be either proteins, polysaccharides, lipids or nucleic acids, while immunogens are usually proteins and large polysaccharides. In Fig. 30.1, a pirate is compared with a pathogen.

Let's assume that a bacteria is a pirate. How can you identify a pirate? Have a look at the diagram below

All the numbered items can help in the identifition of a pirate. These are like antigens present on the surface of a pathogen.

FIGURE 30.1 Antigens of a pathogen.

Sweet Biochemistry. DOI: https://doi.org/10.1016/B978-0-443-15348-8.00020-X

The factors determining immunogenicity of a molecule are its molecule nature, dose, route of entry, additional support of molecules like adjuvants and genetic makeup of host. Nature of antigen like protein content, size and solubility are important as lipids and nucleic acids are weak antigens. Dose determines the antibodies produced against the antigen. Low dose of antigen is handled by low amount of antibodies having high affinity and high specificity. When antigen is present in moderate amounts, higher concentration of antibodies with mixed affinity and broad specificity is reached. At high levels, antigens induce tolerance which means immunity ignores the antigen. Route of entry is related to the region involved in immune response. For example, regional lymph nodes are activated in intradermal, intramuscular and subcutaneous routes. Spleen participates in immune response during intravenous entry of antigen. When antigen comes via mouth or nose, Peyer's patches and bronchial lymphoid tissue are employed. Adjuvants are added to strengthen the effect of antigen. Antigen is responded differently by different people due to species variations and individual variations.

The specific antigenic determinants capable of binding individual antibodies are labelled as epitopes. So on one antigen, multiple epitopes can be present as shown in Fig. 30.2. Antigens may be categorised into thymus-dependent

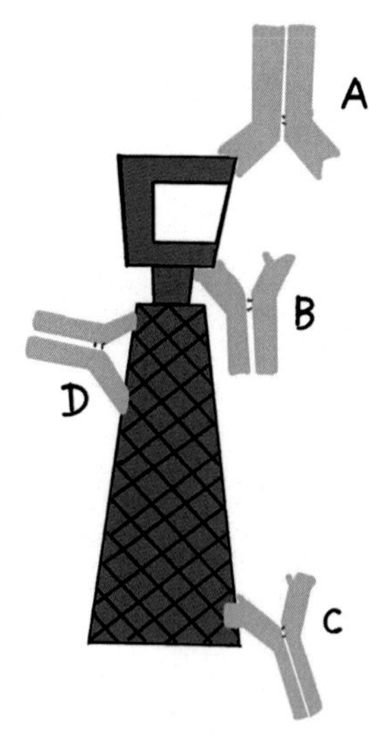

Here you should notice that antibodies are binding at specific different parts of an antigen (drawn as a sword). These parts that are complementary to antibody's binding sites are called epitopes.

FIGURE 30.2 Epitope of A.

and thymus-independent antigens. As the name suggests, the former antigens including proteins and foreign red cells need T-cell involvement to initiate antibody production. Second group of antigens are able to activate specific B-lymphocytes directly by crosslinking antigen receptor on B cell surface. Examples of thymus-independent antigen are bacterial polysaccharides and endotoxins.

Hapten

Molecules like peptides, carbohydrates, nucleic acids, lipids and small organic molecules having low molecular size behave as hapten. Hapten is an incomplete antigen which means that the molecule can attach with antibodies specifically but is unable to induce antibody production itself. By attaching a large size carrier protein, sufficient number of epitopes are collected for activating CD4 + T helper cells. Haptens are commonly electrophilic in nature and bind nucleophilic residues on endogenous proteins covalently. Penicillin antibiotic, metals like nickel and many chemicals behave as hapten.

Superantigens

In normal immune response, almost 0.0001% T cells are activated. But if an antigen activates approximately 30% of T cells, what it should be called? It is called a superantigen. Any antigen is recognised by the receptor present in the surface of immune cell. This unconventional antigen binds the receptor of T cell or B cell very smartly outside the recognition site known as complementary-determining regions (CDRs). T cell superantigens like enterotoxin of *Staphylococcus aureus* and *Streptococcus pyogenes* crosslink the T cell receptor and MHC class II molecules available on antigen-presenting cell. This causes recruitment of lymphocytes along with massive release of cytokines, T-cell apoptosis. On the other hand, B cell superantigens bind antibody on B cell outside the CDR. The hyperstimulation results in the activation of B cell receptor (BCR)-dependent signalling. Due to this, many BCRs are downregulated and CD receptors are unregulated. All these promote release of proapoptotic signals and finally mitochondria permeabilisation and cell death. Examples of B cell superantigen are Staphylococcal Protein A and Streptococcal Protein G.

Neoantigens

Neoantigens are tumour-specific non-autologous antigens produced by tumour cells as a result of non-synonymous mutations in the tumour cells. These are basically new antigens which have not been exposed to immune cells and therefore are attacked. Other causes of expression of neoantigens are post-viral infections, alternative gene splicing and gene rearrangements. Being highly immunogenic and having significant affinity towards MHC, neoantigens allow easy identification of cancer cells to CD4 + and CD8 + T cells. Due to specific association of neoantigens with a tumour, immunotherapy can be targeted against these unlike tumour-associated antigens that are also expressed in normal cells.

Autoantigen

Autoantigens are endogenous molecules that are mislabelled as foreign molecule and attacked by immune system. There are some common properties of such antigens like autoantigens are mostly conserved during evolution means these are some super important molecules (present in functional regions) which may be similar in a bacteria and human. Thus, infections can initiate the process of autoimmunity. In process of attacking the infectious agent, our immune system targets our endogenous antigen. Changes may occur after translation or due to some somatic mutation in the structure of a protein. Such molecules are usually acted upon by caspases. Autoantigens can bind immune receptors. Mechanisms forwarded for explaining autoimmunity include failure to remove autoreactive lymphocytes, molecular mimicry, abnormal expression of normal protein for example encountering hidden self-antigen, epitope spreading and polyclonal lymphocyte activation. Example of autoantigens include p53, dsDNA, ribosomal protein P etc.

Adjuvants

Immune response against an antigen can be enhanced in terms of magnitude and duration by molecules referred as adjuvants. Like adjectives that add some quality to a noun, adjuvants increase immunogenicity of an antigen. Alum, introduced as adjuvant in 1920, is one of the most commonly used molecules. An antigen if modified chemically with short peptides of repeated phosphoserine, the binding to alum is increased. Now this adjuvant is responded by antibodies and CD4 + T helper cells. The underlying mechanism is the stimulation of adaptive immunity by tissue damage which causes uric acid-mediated activation of inflammatory dendritic cells. Another adjuvant is MF59 used in influenza vaccine.

Visual mnemonics for important types of antigens have been provided in Fig. 30.3.

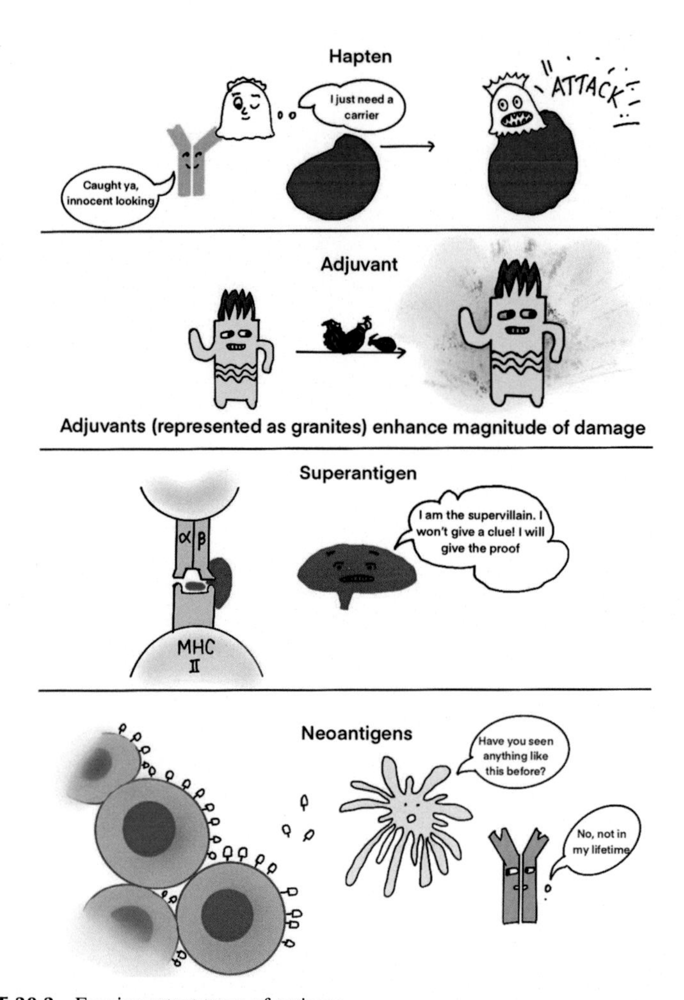

FIGURE 30.3 Few important types of antigens.

Chapter 31

Major histocompatibility complex

Major histocompatibility complex (MHC) is also called histocompatibility antigen and human leucocyte antigen (HLA). 'Histo' is related to tissues, and compatibility is literally defined as a state in which two things can coexist without conflict. Therefore, if an organ is transplanted to a person without proper HLA matching, a severe vigorous immune reaction is observed. Considerable genetic polymorphism is observed along with multiple alleles at each locus. The benefit of variations is the ability to handle emerging new infectious pathogens. Due to genetic polymorphism, in terms of HLA we are very different from each other and only blood-related relatives have matching HLA. Hence, these molecules are very important but have another very pivotal role in immunity.

HLA presents the processed peptides of antigen on the cell surface to the T cells for stimulating immune response. In simple words, HLA is acting as a middle man or linker between the antigen and T cells. This connection is desperately needed by T cell receptor, and antigen is recognised only as a complex of antigenic peptide plus self-MHC. In this way, MHC restricts the dual recognition of antigen by T cell.

Now the question arise, what is HLA structure-wise? This rhyming confusion is noted many times among students who finds MHC topic difficult. So to clearly state, the MHC is cell surface glycoprotein and it is obvious that a protein comes from a gene; hence, there must be MHC regions in chromosomes.

The categorisation of human leucocyte antigen forms three classes I, II and III

MHC class I antigen is composed of variable alpha heavy chain of 45 kDa and a constant beta 2-microglobulin of 12 kDa. Fig. 31.1 depicts a cartoon version of MHC class I and class II structures.

MHC class I and II have similar overall fold. In class I, the binding platform is formed by two domains of one alpha chain, while in class II, both alpha and beta chains contribute. This groove formed can accommodate a

Sweet Biochemistry. DOI: https://doi.org/10.1016/B978-0-443-15348-8.00027-2

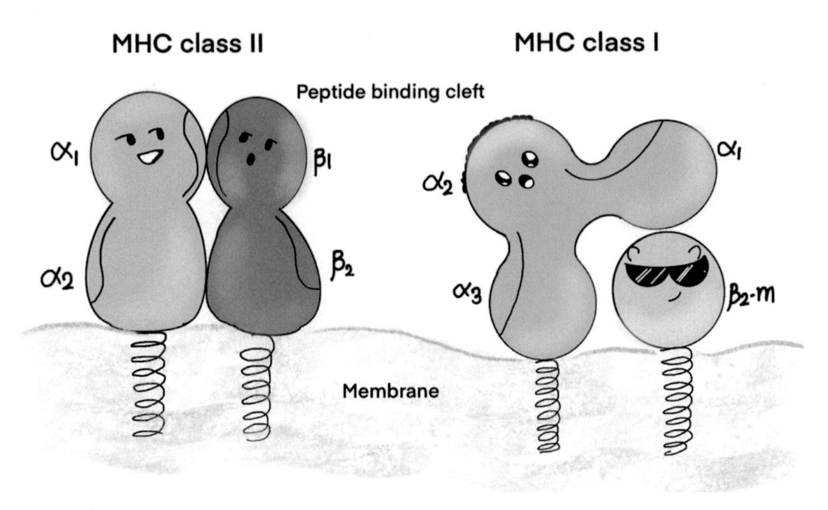

FIGURE 31.1 MHC class I and II structure.

Class I & II MHC details learning

Class I composed of : 1 big chain, 1 small chain and Class 2 by : 2 equal chains (almost)

CD cell relation - an easy trick
1×8= 8 so MHC class I presents to CD8+ cell.
2×4= 8 so MHC class II presents to CD4+ cell.

Class I presents Endogenous antigen's peptides while Class II takes Exogenous. Priority is given to the antigen that is already inside the cell boundary.

FIGURE 31.2 MHC remembering.

peptide chain in between. Approximately nine amino acid long peptide is bound by class I which is expressed on nucleated cells. Fig. 31.2 will help you remember some important points of MHC.

Have a look at Table 31.1 summarising the differences between MHC class I and class II.

MHC class III complex includes the genes encoding proteins participating in complement system like C4, C2 and factor B.

The pathways of antigen presentation by class I and class II are shown in Fig. 31.3.

TABLE 31.1 Differences between class I and class II Major histocompatibility complex (MHC).

	MHC class I	MHC class II
Distribution/ expression	Expressed on all nucleated cell surface	Expressed on antigen-presenting cells like macrophages, dendritic cells and B cells
Composition	Three alpha domains + one beta 2-microglobumin domain	Two alpha + two beta domains
Membrane spanning domain	One alpha domain	One alpha and one beta domains
Size of peptide-binding cleft	Can accommodate peptides with 8–11 residues	Can accommodate peptides with 10–30 residues or more
Encoding genes	Alpha chain is encoded on MHC locus of chromosome 6, while beta chain gene is on chromosome 15. HLA-A, HLA-B and HLA-C genes are involved	HLA-D region on chromosome 6
Antigen handled	Endogenous antigens, viral antigens	Exogenous antigens from pathogens
Antigen-presenting domains	Alpha 1 and alpha 2	Alpha 1 and beta 2
Antigens presented to	Cytotoxic T cells	Helper T cells
Receptor involved	CD8 + receptor on cytotoxic T cells	CD4 + receptor on helper T cells
Antigen processing	Endogenous antigen is broken down into small peptides by proteasome and then transported by shuttle protein called TAP I and TAP II to endoplasmic reticulum. From ER, the antigenic peptides are displayed on surface by MHC-1 proteins to the CD8 + cells	Exogenous antigen is endocytosed and broken down by proteases in the acidic pH of endosome to smaller peptides. In ER, alpha and beta chains of MHC class II proteins combine with an invariant chain polypeptide. This complex is moved to Golgi bodies where it fuses with endosome. Here, the invariant chain is removed by proteases and MHC class II protein bind new antigenic peptides. Then these are exported to the surface

(*Continued*)

TABLE 31.1 (Continued)

	MHC class I	MHC class II
Enzymes involved in peptide generation	Cytosolic proteasome	Endosomal and lysosomal proteases
Variant chain	Absent	Present

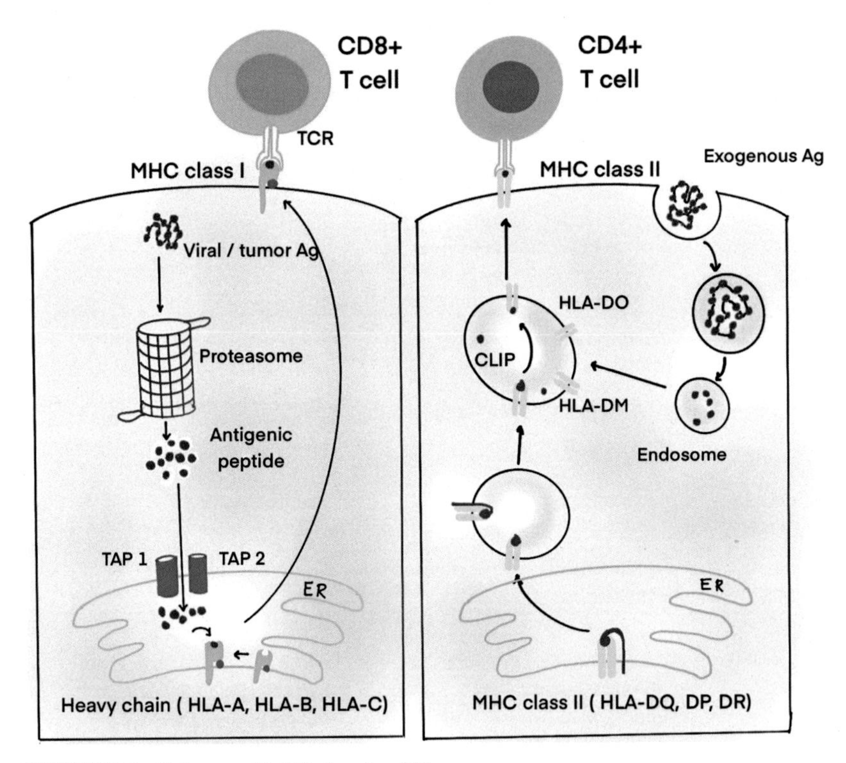

FIGURE 31.3 Pathways of MHC class I and II.

Chapter 32

Vaccines

Vaccine can be defined as any substance that is used to stimulate the body's immune response against one or few diseases. Basically, all types of vaccines present a glimpse of a pathogen or part of a pathogen to the immune system so that it is easy to identify and remove it efficiently on the next exposure. Vaccines are divided into (1) whole pathogen vaccines, (2) subunit vaccines, (3) nucleic acid vaccines and (4) viral-vectored vaccines as classified in Fig. 32.1. Readers would love to see the amazing figure (Fig. 32.2) which displays a simplified trick to recall various types of vaccines.

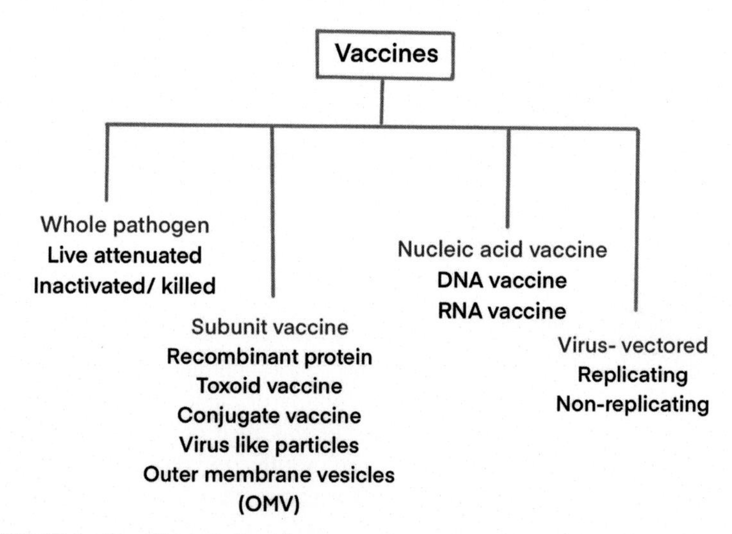

FIGURE 32.1 Classification of vaccines.

Sweet Biochemistry. DOI: https://doi.org/10.1016/B978-0-443-15348-8.00005-3

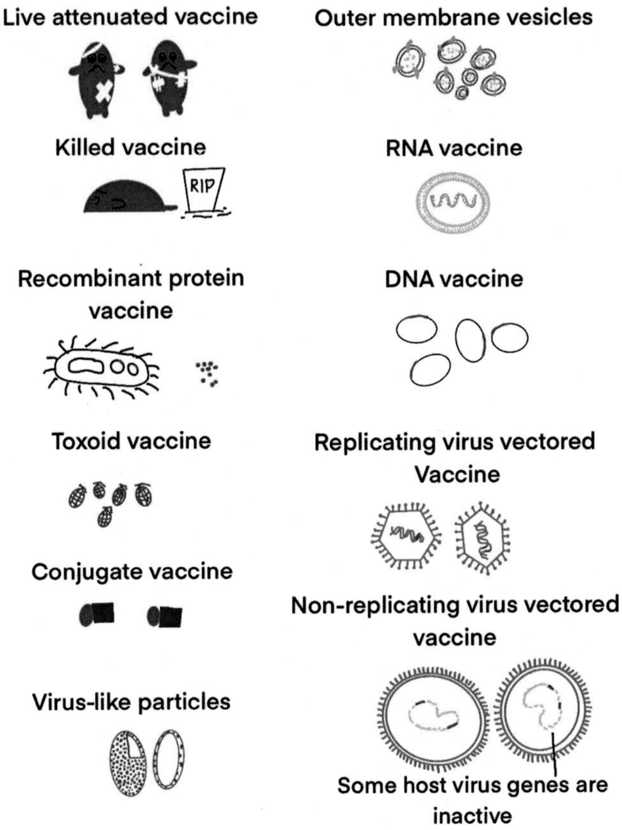

Live attenuated vaccine
Outer membrane vesicles
Killed vaccine
RIP
RNA vaccine
Recombinant protein vaccine
DNA vaccine
Toxoid vaccine
Replicating virus vectored Vaccine
Conjugate vaccine
Non-replicating virus vectored vaccine
Virus-like particles
Some host virus genes are inactive

FIGURE 32.2 Various types of vaccines.

Whole pathogen vaccines

Earlier vaccines used whole pathogen that causes immune response resembling the natural infection. But due to possibility of dangerous effects of disease as well as spread to others, the use of this type of vaccines has been minimised. Pathogen form can be weakened or killed to yield vaccines known as live attenuated vaccines and inactivated vaccines.

1. **Live attenuated vaccine**. If a virus or bacteria is allowed to grow in unfavourable circumstances in culture medium (high or low temperature, pH and poor nutrients) for a long span of time, for example 10 years, the virulence of wild pathogen decreases by mutations. This is known as attenuation. Less virulent forms can also be created due to artificially introduced modifications (genetic/gamma or UV ray irradiation). Live attenuated vaccines lead to lifelong immunity with even one or two doses. There is always the possibility of reverting back to wild pathogen

form; hence, live attenuated vaccine cannot be administered in person with weak immunity. Examples are rotavirus vaccine and MMR vaccine.

2. **Inactivated vaccine**. In this type, pathogen is killed or altered to an extent that replication is impossible. So there is no risk that virus will again come to life and say 'I am back Ha ha ha'. Inactivated pathogen cannot cause illness; therefore, it can be given to people with poor immunity. The immune response generated is however less strong and short term in comparison to live attenuated vaccines. Examples are inactivated polio vaccine, hepatitis A vaccine.

Subunit vaccine

If we do not want to take whole pathogen and rather use its small component like one or more surface antigens, the vaccine is called subunit vaccine. The advantage is that the immune response can identify and target these specific targets. Of course, a pathogen with thousands of antigens will provide better stimulation, but the immunogenicity of subunit vaccines can be increased by adding adjuvants or giving repeated doses and booster doses. Adjuvants are substances which add to the strength and duration of immune response against an antigen.

Recombinant protein vaccine

Important surface antigens of pathogen can be synthesised by other organisms like yeast or bacteria by inserting the genes of pathogen coding that antigens into the DNA of yeast. This fused foreign DNA sequence will be transcribed and translated along with the host DNA forming the antigen molecules by the yeast protein machinery. These surface antigens can be purified and used in vaccine. Examples are hepatitis B vaccine and HPV vaccine.

Toxoid vaccine

Many bacteria produce deadly toxins for attacking and are responsible for predominant effects of disease. If immune system can identify the toxins, disease will be ameliorated. So inactivated forms of toxins are used to stimulate the immune response and remember the toxins. Such vaccines are referred to as toxoids due to the resemblance with toxins. Examples are diphtheria vaccine and tetanus vaccine.

Conjugate vaccine

Conjugation means attaching to something. It is important to remember who is pairing with whom and why? For some bacteria, the polysaccharides on surface are more important than proteins. But the vaccines of polysaccharide molecules did not work well in paediatric age group. Hence to amplify the immunogenicity, the polysaccharide was conjugated with the toxoid protein of diphtheria or tetanus which are highly immunogenic. Examples are children's pneumococcal vaccine and Hib vaccine.

Virus-like particles

As the name suggests, virus-like particles (VLPs) are like the shells of virus and don't have the genetic material that removes the chances of infection. Naturally, this type of structures are made, but structural components of viruses can also self-assemble to form VLPs. On VLPs, more than one copy of one antigen from one pathogen or multiple pathogens can be expressed. VLP can also act as adjuvants. Examples are hepatitis B vaccine and HPV vaccine.

Outer membrane vesicle vaccines

Outer membrane vesicles (OMVs) are a bleb (bubble-like structure) of outer cell wall of bacteria. This noninfectious structure possesses numerous antigens of antigens. OMVs are collected in lab from bacteria and incorporated in vaccine. Selective placement or removal of bacteria antigens from OMVs is possible. Like VLPs, OMVs can also act as adjuvants.

Nucleic acid vaccines

This promising class of vaccines provide genetic information for synthesis of pathogenic antigens in cells for generating immune response in humans. Nucleic acid vaccines are easy to produce and lead to good immune responses. Various types of nucleic acid vaccines are RNA vaccines and DNA vaccines.

1. **RNA vaccines**. RNA vaccines are composed of mRNA coding antigen protein wrapped in a lipid membrane. Lipid membrane protects the mRNA during entry in body and facilitates cell import by membrane fusion. Inside the cell, mRNA gets translated into antigenic proteins which instigate the immune response. mRNA degrades after few days. One advantage is that mRNA is not capable of insertion into human genome. Examples are Pfizer BioNTech and Moderna COVID-19 vaccines.
2. **DNA vaccines**. Foreign DNA insertion is more advantageous as DNA is more stable and don't require initial protection. Electroporation in which low-intensity electronic waves cause cell to uptake DNA is applied to administer DNA vaccine. Examples are yet under research zone.

Viral-vectored vaccines

Innocent viruses can be used to carry pathogenic genetic code which can translate into specific antigens after the infection of virus into body cells.

This mechanism is like whole virus vaccine, but here the virus itself is not harmful. It is just delivering the viral DNA. Viral vectors can be replicating or nonreplicating, depending on the ability to replicate. One dose of replicating viral-vectored vaccines is sufficient to induce immunity. Example of this type is Ebola virus vaccine. Advantage of nonreplicating viral vectors is that there is no chance of causing disease. Example of nonreplicating vaccine is Oxford-AstraZeneca COVID-19.

Exercises

1. Identify the three pyrimidines and write their names.

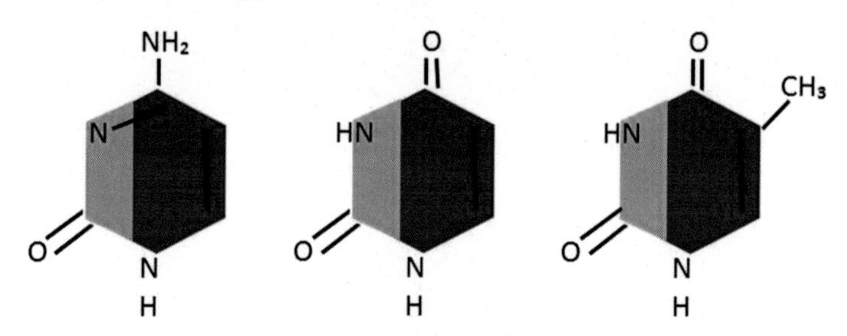

2. Write the names of the intermediates of de novo pathway of purine.

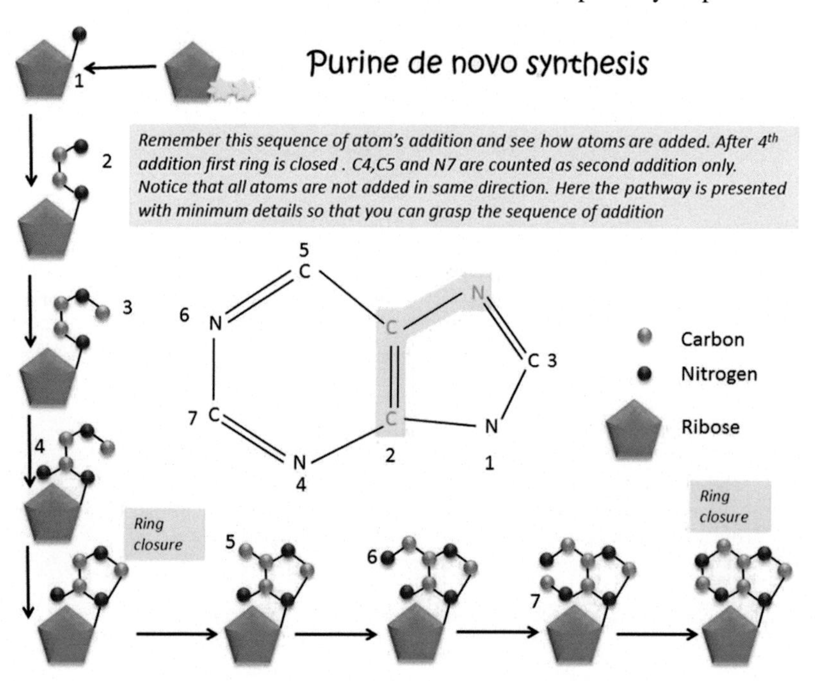

3. Why has this girl kept her hand on her abdomen? Identify the disease and enzyme defect.

4. Write the glycogen storage enzyme defect in empty column.

Type	Story part correlating	Disease	Enzyme defect
0	It's a Great Story	Glycogen Synthase defect	
1a	Von (one) Gierk was a Great 6 pack abs warrier	Von Gierke's disease	
1b	and transported		
II	A pompe with acid maltase	Pompe's disease	
III	He adored Cori-a debrancher	Cori's disease	
IV	While he was a brancher from andery	Anderson's disease	
V	He fought with McArdle with large muscle phoscle	McArdle syndrome	
VI	To win Her as life partner	Her's disease	
VII	On the day of Tarui, with blood (RBC)-stained muscles and one PFK sword	Tarui's disease	

(*Continued*)

(Continued)

Type	Story part correlating	Disease	Enzyme defect
VIII	He suffered 8 liver powerful kuts (cuts)		
IX	And 9 muscle phoscle kuts (cuts)		
X	Finally impressed, she married him in the camp of deproka		

5. See the Queen Bee's house and draw actual cholesterol structure.

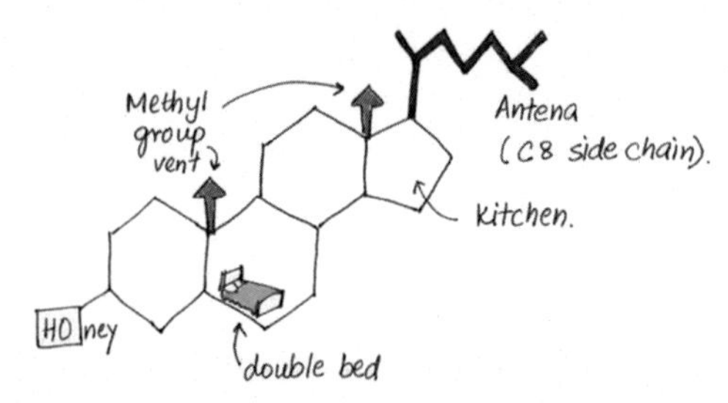

6. See the story of etc from Chapter 3 and write below the etc components correlating with them.

7. See the mnemonic of Glycolysis and then write the reaction names in the box below.

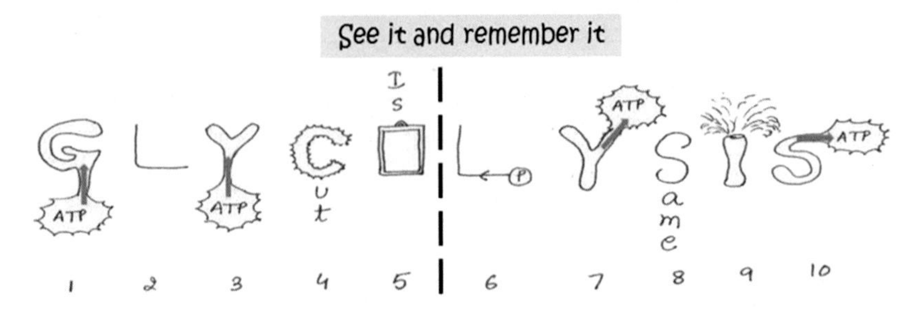

8. Write in the table steps of cholesterol synthesis which correlates with the story.

What bee does	What correlates in cholesterol pathway
2C	
4C	
6C	
5C	
10C	
15C	
30C	
Ouspicious (auspicious) board at the entrance of house	
She rearranges the walls	
She discards the extra hangings below floor	
Next she moved her bed	
She corrected her TV antenna	

9. Identify the disease and enzyme defect. Can you write other two person's names who were standing with him?

10. Write the sources of atoms of purine ring.

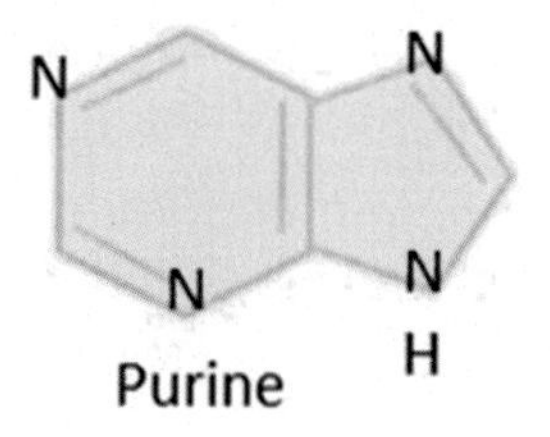

Further reading

Berg, J., Tymoczko, J.L., Stryer, L., 2015. Biochemistry, eighth ed. WH Freeman and Company, New York, United States.

Bhagavan, N.V., 2012. Medical biochemistry, fourth ed. Academic Press, Elsevier, California, United States. Available from: https://rarediseases.org/rare-diseases.

Devlin, T.M., 2010. Textbook of biochemistry with clinical correlations, seventh ed. John Wiley & Sons, New York, United States.

Gupta, S.K., 2017. Essentials of immunology, second ed. Arya Publications, New Delhi.

Rodwell, V.W., Bender, D., Botham, K.M., Kennelly, P.J., Weil, P.A., 2014. Harper's illustrated biochemistry, thirteenth ed. McGraw-Hill Education, New York, United States.

Vasudevan, D.M., Sreekumari, S., Vaidyanathan, K., 2016. Textbook of biochemistry for medical students, eighth ed. Jaypee Brothers Medical Publishers, New Delhi.

Index

Note: Page numbers followed by "*f*" and "*t*" refer to figures and tables, respectively.

A

F

G

H

R

S

T

Printed in the United States
by Baker & Taylor Publisher Services